Presented by

Graham Richards

(Visiting Prof UC Berkeley 1975)

THE INTERNATIONAL SERIES OF MONOGRAPHS ON CHEMISTRY

SPIN-ORBIT COUPLING IN MOLECULES

BY

W. G. RICHARDS
H. P. TRIVEDI
AND
D. L. COOPER

CLARENDON PRESS · OXFORD · 1981

Oxford University Press, Walton Street, Oxford OX2 6DP

OXFORD LONDON GLASGOW
NEW YORK TORONTO MELBOURNE WELLINGTON
KUALA LUMPUR SINGAPORE JAKARTA HONG KONG TOKYO
DELHI BOMBAY CALCUTTA MADRAS KARACHI
NAIROBI DAR ES SALAAM CAPE TOWN

Published in the United States by Oxford University Press, Inc., New York

British Library Cataloguing in Publication Data

Richards, W. G.
Spin–orbit coupling in molecules. –
(The international series of monographs on chemistry)
1. Molecular orbitals – Mathematics
2. Quantum chemistry – Mathematics
I. Title II. Trivedi, H. P.
II. Cooper, D. L.
541.2'8 QD461

ISBN 0-19-855614-4

Typeset by Anne Joshua Associates, Oxford
Printed in Great Britain
at the University Press, Oxford
by Eric Buckley
Printer to the University

PREFACE

SPIN-ORBIT coupling is the almost universal symmetry breaker which allows the mixing of energy levels which would otherwise be orthogonal. The influence of spin–orbit coupling is thus evident in a wide range of scientific disciplines extending from nuclear physics through solid-state physics to organic chemistry and even astrophysics. It is however in the realm of atomic and molecular physics that the effects of introducing spin–orbit coupling are best understood and that the theory stands up best to quantitative examination.

The present monograph concentrates on this area where both experiment and theoretical calculation are at their most accurate and of comparable precision. It thus deals with atoms by way of introduction but the bulk of the text is devoted to spin–orbit coupling effects in diatomic molecules, where the experimental data are provided by high-resolution spectroscopy.

Although spin–orbit splittings are observed in the spectra of transition metal ions and off-diagonal spin–orbit effects are of considerable importance in the intersystem crossings observed in organic photochemistry, we have chosen to ignore these more crudely observed and less well understood phenomena and to limit consideration to the small gas-phase species.

Despite the importance of spin–orbit coupling and the notable successes of theoretical calculations, of both diagonal and off-diagonal effects, there has been relatively little published work involving computations. This is particularly striking when compared with the vast literature on *ab initio* calculation of molecular wavefunctions. There are two reasons for this, the first being the lack of a suitable introductory text and the second the absence of freely available computer programs. We hope to have provided for both these needs, including in the text an up-to-date version of the notation on symbols and symmetries, which complicate the topic, and as appendices and a microfiche, details of computer programs which permit the calculation of all the topics discussed in the main body of the book. A bibliography of the relevant theoretical papers and some experimental papers published up to the end of 1979 is also included. At the end of each chapter, key references are given under the heading 'Further reading'.

It is our hope that we can encourage more workers to enter this particular field where there is an abundance of problems and that we can promote the extension of work on spin–orbit coupling effects into the territory of polyatomic molecules and ions.

We owe more than the customary acknowledgement to the line of graduate students whose work is summarized here. In particular we are grateful to Timothy Walker who started so many of the hares which were subsequently chased by Anthony Hall, Reg Hinkley, John Raftery, Bob Hammersley, Ian Wilson, and Elizabeth Gold. Our thanks are also due to Judith Adam who produced a clear typescript from a daunting manuscript.

A large proportion of the mathematical equations in this monograph were drawn on the III FR 80 film recorder of the Rutherford Laboratory using programs developed for this purpose called SOFTY. Most of the characters were ultimately set using the subroutine SYMBOL due to N. Wolcott of the US National Bureau of Standards and made available at the Rutherford Laboratory by K. M. Crennell. The character definitions are those of A. C. Hershey and the graphics system was SMOG.

We are grateful to the Science Research Council for computing facilities. The help of many of the staff at the Rutherford Laboratory has been invaluable, particular the operators of the two IBM 360/195 computers and IBM 3032 on which SOFTY was run and the program advisers. A full list of those who helped would make this a very long section, but one small group does deserve special mention: they are the four operators of the III FR 80.

Oxford
June 1980

W.G.R.
H.P.T.
D.L.C.

CONTENTS

INTERNATIONAL SERIES OF
MONOGRAPHS ON CHEMISTRY

J. D. Lambert: *Vibrational and rotational relaxation in gases*
N. G. Parsonage and L. A. K. Staveley: *Disorder in crystals*
G. C. Maitland, M. Rigby, E. B. Smith, and W. A. Wakeham: *Intermolecular forces: their origin and determination*
W. G. Richards, H. P. Trivedi, and D. L. Cooper: *Spin–orbit coupling in molecules*
C. F. Cullis and M. M. Hirschler: *The combustion of organic polymers*
R. T. Bailey, A. M. North, and R. A. Pethrick: *Molecular motion in high polymers*

1

SPIN–ORBIT COUPLING IN ATOMS

1.1. Introduction

The first encounter with spin–orbit coupling comes at a very elementary level, in the atomic spectrum of the alkali metals.

The Russell–Saunders coupling scheme suggests that the ground and first excited electronic states of the lithium atom should be ^2S and ^2P arising from electronic configurations $1s^22s$ and $1s^22p$ respectively. The energy difference between the two states results from the different spatial distributions of 2s and 2p electrons which leads to different electrostatic repulsion between the electrons. However, as is well known, the electronic transition between these states gives rise not to one but two lines in the atomic spectrum. The famous sodium D lines are particularly easy to observe.

The fact that two transitions are seen (Fig. 1.1) is due to the splitting of the ^2P energy level into sublevels $^2P_{1/2}$ and $^2P_{3/2}$. The origin of the small splitting is spin–orbit coupling, which is a magnetic effect. In very much simplified terms it may be understood by considering the spinning electron acting as a magnet which then interacts with

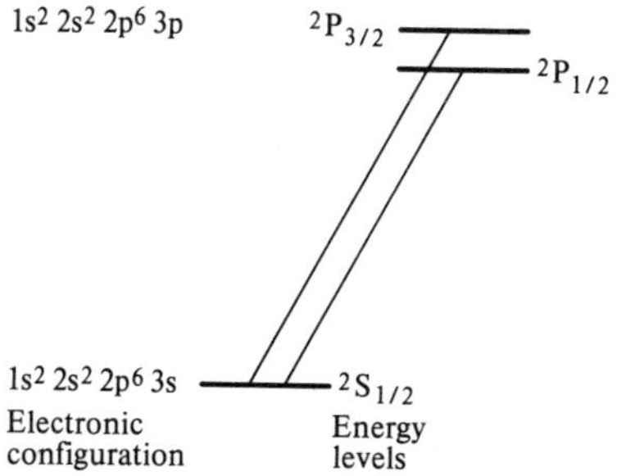

Fig. 1.1. Transitions responsible for the D lines of the sodium atom spectrum showing the ^2P levels split by spin–orbit coupling.

the magnetic field created by its own orbital motion round the nucleus. Such an effect would exist in a classical situation. However, at the atomic level where the effects are quantum mechanical, there is a limitation on the 'orientation' of the effective magnet which is related to the total angular momentum of the system.

Very approximately, as we shall see, the splitting depends on the integral

$$\int \psi \frac{Z_{\text{eff}}}{r^3} \psi \; d\tau \qquad (1.1)$$

where ψ is the wave function, Z_{eff} is an effective nuclear charge of the screened nucleus, and r is the distance of the electron from that nucleus. Since the average distance of an electron from the nucleus is also roughly inversely proportional to the nuclear charge the actual splitting in the ^{2}P level of the alkali metals might be expected to be approximately proportional to Z^4. The observed spin–orbit splittings are shown in Fig. 1.2.

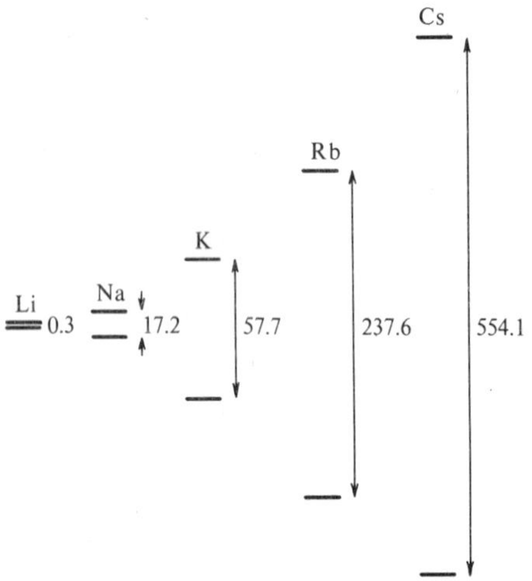

Fig. 1.2. Splittings of the $^2P_{3/2}$ and $^2P_{1/2}$ levels of the alkali metals (cm^{-1}).

This high dependence of the spin–orbit splitting on Z may be appreciated if one realizes that the magnetic interactions of a spinning charge e rotating round a nucleus of charge Z is equivalent to having the nuclear charge Z rotating round the electron; the change being merely a shift of coordinate system. Clearly when viewed in this frame the magnetic effects are likely to be highly dependent on the nuclear charge Z. One minor confusion which does arise in this coordinate change is the introduction of a factor of 2. The factor comes into the problem just as it does in simple questions as to how many times the moon rotates as it goes round the Earth: it all depends on whether you view the problem from the Earth or the moon. This effect, called the Thomas precession, is the origin of factors of 2 which are frequently found in spin–orbit coupling problems and mistakes and confusion about that bedevil some aspects of the topic.

1.2. The qualitative treatment of spin–orbit coupling

'Spin' is of course a relativistic effect, but the classical picture does none the less help in the understanding of spin–orbit coupling. The interaction couples together the spin (**S**) and orbital (**L**) vectors, and reduces the degeneracy of spectroscopic terms with a given L and S, but a different value of J, the total angular momentum. The interaction of the electron's magnetic moment μ with a field **B** is given by $-\mu\mathbf{B}$ and, in the absence of an external field, **B** is proportional to the vector product of the nuclear field and the velocity of the electron, i.e.

$$\mathbf{B} = (\mathbf{E}\times\mathbf{v})/c.$$

Now $\boldsymbol{\mu} = -(eh/2mc)\boldsymbol{\sigma}$ and $\mathbf{v} = \mathbf{p}/m$,

so that the interaction will be

$$(eh/2m^2c^2)\boldsymbol{\sigma}\cdot(\mathbf{E}\times\mathbf{p})$$

and for a central field

$$\mathbf{E} = \frac{1}{r}\frac{\partial V}{\partial r}\,\mathbf{r}$$

with V being the potential.

Writing $\mathbf{s} = h\boldsymbol{\sigma}/2$ and $\mathbf{l} = \mathbf{r}\times\mathbf{p}$ the expression for the interaction becomes

$$(e^2/m^2c^2)\left(\frac{1}{r}\frac{\partial V}{\partial r}\right)\mathbf{l}\cdot\mathbf{s}.$$

We have considered the nucleus to be at rest, and so this is an over-estimate by a factor of 2 – the Thomas precession mentioned earlier.

Since spin is a relativistic effect, it is more correct to start with Dirac's relativistic equation for the electron. The reduction of the Dirac equation to non-relativistic form and the use of atomic units gives the spin–orbit interaction energy for a one-electron atom as

$$\begin{aligned} H_{SO} &= \frac{1}{2m^2c^2}\frac{1}{r}\frac{\partial V}{\partial r}\,\mathbf{l}\cdot\mathbf{s} \\ H_{SO} &= \frac{\alpha^2}{2}\frac{1}{r}\frac{\partial V}{\partial r}\,\mathbf{l}\cdot\mathbf{s}. \end{aligned} \tag{1.2}$$

where V is the Coulomb potential due to the nucleus and α the fine structure constant. This may be rewritten as $\zeta\mathbf{l.s}$ and this provides a definition of the spin–orbit coupling constant ζ for a one-electron atom. In such a case $V = -Z/r$ and so $\partial V/\partial r = Z/r^2$.

The generalization of this result to many electron atoms is far from straightforward but following Condon and Shortley (see Further reading) it is normal to reinterpret V as the potential due to the nucleus and to all the other electrons and the result is then summed over all the outer electrons, giving the spin–orbit hamiltonian as

$$\begin{aligned} H_{SO} &= \sum_i \frac{\alpha^2}{2}\frac{1}{r_i}\frac{\partial V}{r_i}\,\mathbf{L}\cdot\mathbf{S} \\ &= \sum_i \zeta(r_i)\,\mathbf{l}_i\cdot\mathbf{s}_i. \end{aligned} \tag{1.3}$$

In qualitative treatments of spin–orbit coupling the exact nature of the potential V is not explored further.

This spin–orbit hamiltonian does not commute with the operators S_z, L_z, S^2 or L^2, but only with J^2 and J_z. These latter pairs are related to the others by

$$J_z = L_z + S_z$$

and its analogues and by

$$J^2 = J_x^2 + J_y^2 + J_z^2$$

If we have a function ψ which is an eigenfunction of these operators

$$J_z\psi = M_J\psi$$

$$M_J = M_L + M_S$$

$$J^2\psi = J(J+1)\psi$$

$$-J \leqq M_J \leqq J$$

In order to calculate matrix elements of this total hamiltonian including H_{SO} is it convenient to use eigenfunctions of J^2.

1.2.1. *Example: the configuration sp*

We can characterize the ^3P eigenfunctions of this configuration by the values of M_L and M_S (Table 1.1).

Table 1.1

Determinantal wavefunction	M_L	M_S	$^3P_{M_L,M_S}$	M_J
$\lvert sp_+\rvert$	1	1	$^3P_{1,1}$	2
$1/\sqrt{2}\{\lvert s\bar{p}_+\rvert + \lvert \bar{s}p_+\rvert\}$	1	0	$^3P_{1,0}$	1
$\lvert \bar{s}\bar{p}_+\rvert$	1	-1	$^3P_{1,-1}$	0
$\lvert sp_0\rvert$	0	$+1$	$^3P_{0,1}$	1
$1/\sqrt{2}\{\lvert s\bar{p}_0\rvert + \lvert \bar{s}p_0\rvert\}$	0	0	$^3P_{0,0}$	0
$\lvert \bar{s}\bar{p}_0\rvert$	0	-1	$^3P_{0,-1}$	-1
$\lvert sp_-\rvert$	-1	$+1$	$^3P_{-1,1}$	0
$1/\sqrt{2}\{\lvert s\bar{p}_-\rvert + \lvert \bar{s}p_-\rvert\}$	-1	0	$^3P_{-1,0}$	-1
$\lvert \bar{s}\bar{p}_-\rvert$	-1	-1	$^3P_{-1,-1}$	-2

From Table 1.1 one can see that there will be a state with J equal to the maximum value of M_J, i.e. $J = M_J = 2$ called $^3P_2$ for which M_J can be 2, 1, 0, -1, -2. There are also the states $^3P_1$ (with $M_J =$ 1, 0, -1) and $^3P_0$ ($M_J = 0$ only).

These states can be found from the state with $M_J = 2$ of $^3P_2$ by applying the step-down operator

$$J^- = L^- + S^-,$$

remembering that

$$L^{-}(P_m) = \{l(l+1)-m(m-1)\}^{\frac{1}{2}}(P_{m-1})$$

Thus,

$$J^{-}(sp_{+}) = \sqrt{2}|sp_0| + |s\bar{p}_{+}| + |\bar{s}p_{+}|$$

$$= \sqrt{2}\{{}^3P_{0,1} + {}^3P_{1,0}\}.$$

Then repeating the step-down operation we may obtain from ${}^3P_2$, ($M_J = 1$) the component of ${}^3P_2$ ($M_J = 0$). Also the function ${}^3P_1$ ($M_J = 1$) can be obtained by using the condition that it is orthogonal to ${}^3P_2$ ($M_J = 1$).

Complete application of this procedure yields the resulting functions

${}^3P_2$	$M_J = 2$	${}^3P_{1,1}$
	$M_J = 1$	$\frac{1}{\sqrt{2}}[{}^3P_{1,0} + {}^3P_{0,1}]$
	$M_J = 0$	$\frac{1}{\sqrt{6}}[2{}^3P_{0,0} + {}^3P_{1,-1} + {}^3P_{-1,1}]$
${}^3P_1$	$M_J = 1$	$\frac{1}{\sqrt{2}}[{}^3P_{1,0} - {}^3P_{0,1}]$
	$M_J = 0$	$\frac{1}{\sqrt{2}}[{}^3P_{1,-1} - {}^3P_{-1,1}]$
${}^3P_0$	$M_J = 0$	$\frac{1}{\sqrt{3}}[{}^3P_{1,-1} - {}^3P_{0,0} + {}^3P_{-1,1}]$.

As is explained in many elementary textbooks these functions are in practice obtained using Clebsch–Gordan or $3j$ coefficients (see Further reading).

1.3. Calculation of matrix elements of the spin–orbit hamiltonian

1.3.1. *Method 1*

The problem can be tackled directly by using Slater's rules for taking matrix elements and by assuming that the operator is a one-electron operator as is the case for single-electron atoms.

The energy resulting from electrostatic interactions depends on L and S and so is identical for all the levels of a given term such as a ^3P state.

We have

$$\mathbf{l}\cdot\mathbf{s} = l_x \cdot s_x + l_y \cdot s_y + l_z \cdot s_z$$

$$= l_z s_z + \tfrac{1}{2}[(l_x + il_y)(s_x - is_y) + (l_x - il_y)(s_x + is_y)$$

$$= l_z s_z + \tfrac{1}{2} l^+ s^- + \tfrac{1}{2} l^- s^+.$$

Thus,

$$\langle {}^3P_2 | H_{SO} | {}^3P_2 \rangle = \langle s | \zeta(r_1)\, l(1)\, s(1) | s \rangle + \langle p_+ | \zeta(r_2) l(2) s(2) | p_+ \rangle$$

In the diagonal element only the contribution due to $l_z s_z$ remains:

$$l_z(1)\, s_z(1)\, p_+(1) = +\tfrac{1}{2} p_+(1),$$

and so

$$\langle {}^3P_2 | H_{SO} | {}^3P_2 \rangle = \tfrac{1}{2} \langle p | \zeta(r) | p \rangle = \tfrac{1}{2}\zeta$$

with ζ being a radial integral.

$$\langle {}^3P_1 | H_{SO} | {}^3P_1 \rangle = \tfrac{1}{2}[\langle {}^3P_{1,0} | H_{SO} | {}^3P_{1,0} \rangle + \langle {}^3P_{0,1} | H_{SO} | {}^3P_{0,1} \rangle$$

$$+ 2\langle {}^3P_{1,0} | H_{SO} | {}^3P_{0,1} \rangle]$$

$$= \tfrac{1}{2}\{-\tfrac{1}{2}\zeta - \tfrac{1}{2}\zeta + 2 \times 0\}$$

$$= -\zeta/2.$$

For $\langle {}^3P_0 | H_{SO} | {}^3P_0 \rangle$ there are also off-diagonal terms such as for example $\langle {}^3P_{1,1} | H_{SO} | {}^3P_{0,0} \rangle$ with a contribution from

$$\tfrac{1}{2}(l^+ s^-) | \bar{s} p_0 | = \sqrt{2} | \bar{s} \bar{p}_+ |$$

from which

$$\langle {}^3P_{1,-1} | H_{SO} | {}^3P_{00} \rangle = \tfrac{1}{2} \langle \bar{s}\bar{p}_+ | \zeta(r) | \bar{s}\bar{p}_+ \rangle = \zeta/2.$$

Similarly

$$\langle {}^3P_{-1,1}|H_{SO}|{}^3P_{0,0}\rangle = \tfrac{1}{2}\zeta.$$

Thus

$$\langle {}^3P_{0,0}|H_{SO}|{}^3P_{00}\rangle = \tfrac{1}{3}[-\zeta/2-\zeta/2-2\zeta/2-2\zeta/2] = -\zeta.$$

The energy separation between ${}^3P_2$ and ${}^3P_1$ is thus twice that between ${}^3P_1$ and ${}^3P_0$.

1.3.2. *Method 2*

The above, rather long-winded method can be avoided by utilizing the angular parts of the function. The diagonal matrix element of the spin–orbit operator for an eigenfunction of J^2 may be written as

$$\langle L\ S\ J\ M_J|\sum_i \zeta(r_i)\mathbf{l}_i\cdot\mathbf{s}_i|L\ S\ J\ M_J\rangle$$

and for the electron labelled n, we have

$$\langle L\ S\ J\ M_J|\zeta(r_n)\mathbf{l}_n\cdot\mathbf{s}_n|L\ S\ J\ M_J\rangle.$$

The total matrix element is formed by summing over n.

However, the spin–orbit operator is the product of spin and spatial parts, so that the above may be rewritten using Racah's formula as:

$$(-1)^{S+L+J}\begin{Bmatrix} S & L & J \\ L & S & l \end{Bmatrix}\ \langle L\ S||\zeta(r_n)\mathbf{l}_n||L\ S\rangle.$$

Here $\langle LS\|\ \|LS\rangle$ is a reduced matrix element independent of the quantum numbers M_L and M_S and only depends on L and S. The term

$$\begin{Bmatrix} S & L & J \\ L & S & l \end{Bmatrix}$$

is the 6_j symbol which describes the coupling of the vectors L and S to form J. The value of this 6_j symbol is

$$\frac{J(J+1)-L(L+1)-S(S+1)}{\surd\{S(2S+1)(2S+2)L(2L+1)(2S+1)\}}.$$

When we add the dependence on J we have then,

$$\langle L\ \ S\ \ J|\mathrm{H}_{\mathrm{SO}}|L\ \ S\ \ J\rangle = \tfrac{1}{2}A_{L,S}[J(J+1)-L(L+1)-S(S+1)].$$

Similarly for the state $J-1$

$$\langle L\ \ S\ \ J|\mathrm{H}_{\mathrm{SO}}|L\ \ S\ \ J-1\rangle = \tfrac{1}{2}A_{L,S}[(J-1)J-L(L+1)-S(S+1)]$$

whence the energy difference is

$$E_J-E_{J-1} = \tfrac{1}{2}A_{L,S}[J(J+1)-J(J-1)] = A_{L,S}J$$

This is the Landé interval rule: hyperfine spacings are proportional to J.

The constant A is positive for electronic configurations with less than half-filled shells and negative if the electronic shell is more than half-filled (i.e. we may consider holes in the electronic shell).

In our example (the p^2 configuration), the energy diagram is thus as in Fig. 1.3. This picture is valid for Russell–Saunders coupling scheme cases where the fine structure separation due to spin–orbit coupling is small. The scheme neglects spin–spin interactions which give rise to deviations from Landé's rule.

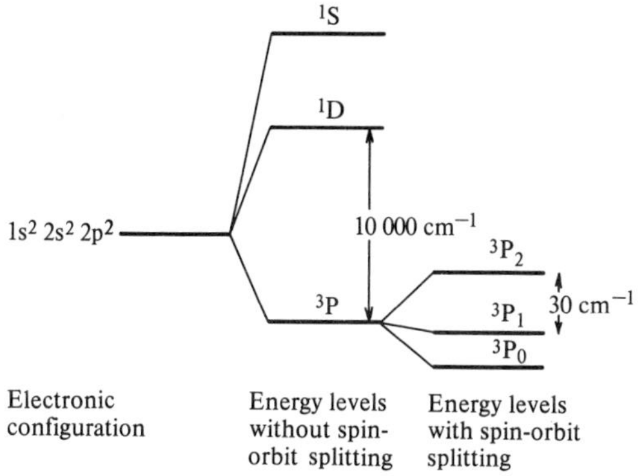

Fig. 1.3. Energy level diagram for the configuration p^2.

1.4. The quantitative treatment of spin–orbit coupling

Although this book deals with the calculation of spin–orbit coupling constants in molecules, it is necessary to go into atomic calculations in some detail. Molecular wavefunctions are usually expressed in terms of atomic orbitals so that much of the analysis of atomic systems will carry over directly. Some of the lengthy formulae derived for atoms may be used without change, even though the actual integrals which have to be evaluated are made more difficult.

1.4.1. *The nature of the operator*

The generalization of the one-electron atomic expression to poly-electronic atoms was achieved by Condon and Shortley by interpreting the potential as being due to the nucleus and to all the electrons in closed shells,

$$V = \frac{\alpha^2}{2}\left(\frac{1}{r}\frac{\partial V}{\partial r}\right)\sum_i \mathbf{l}_i\cdot\mathbf{s}_i.$$

This expression does not always lead to a correct explanation of the fine structure of light atoms which may depart radically from the Landé interval rule. It is also far from clear whether exchange terms such as occur in energy expressions derived using Hartree–Fock wave-functions should be included.

These problems may be overcome by including terms involving the spin of one electron with the orbit of another and spin–spin interactions. The spin–other-orbit term provides a shielding factor since the potential in which the outer electrons move may be written as

$$V = \frac{Z}{r} - \sum_{i \neq j}\frac{1}{r_{ij}}.$$

This leads to a hamiltonian

$$\mathrm{H_{so}} = \frac{\alpha^2}{2}\left\{\sum_i \frac{Z}{r_i^3}(\mathbf{l}_i\cdot\mathbf{s}_i) - \sum_{i \neq j}\frac{1}{r_{ij}^3}(\mathbf{r}_{ij}\times\mathbf{p}_i)\cdot(\mathbf{s}_i+2\mathbf{s}_j)\right\}. \tag{1.4}$$

The second term, the spin–other-orbit interaction, may also be written as

$$-\sum_{i \neq j}\frac{1}{r_{ij}^3}(\mathbf{r}_{ij}\times\mathbf{p}_i)\cdot(\mathbf{s}_i+2\mathbf{s}_j)$$

$$= \sum_{i>j} \frac{1}{r_{ij}^3} \mathbf{r}_{ij} \times (\mathbf{p}_i - \mathbf{p}_j) \cdot \mathbf{s}_j$$

$$= \sum_{i \neq j} \nabla_i \frac{1}{r_{ij}} \times \mathbf{p}_i \cdot (\mathbf{s}_i + 2\mathbf{s}_j).$$

With the shielding thus expressed as a two-electron operator, the exchange terms can be included in exactly the same way as they appear in energy expressions for determinantal Hartree–Fock wave functions.

The hamiltonian (1.3) was derived by Bethe and Salpeter from the Breit equation. It can be simplified by noting that both terms vanish for electrons in closed shells. The two-body interaction does not however vanish between an electron in an open shell and the electrons in closed shells. Using an argument propounded by Elliott for nuclear shells, Blume and Watson have shown that the type of operator given in (1.4) does act like a one-electron operator. Horie has also demonstrated that a part of the spin–other-orbit interaction between the electrons in open shells behaves in this way and should be incorporated into the definition of the spin–orbit coupling constant.

1.4.2. *Matrix elements*

In order to make *ab initio* calculations of spin–orbit coupling constants, matrix elements of the hamiltonian must be taken over atomic orbitals. We must be able to compute

$$\begin{aligned} \lambda LS &= \frac{Z\alpha^2}{2} \sum_a \langle a | \frac{1}{r^3} \mathbf{l} \cdot \mathbf{s} | a \rangle \\ &- \frac{\alpha^2}{2} \sum_{a,b} \{ \langle a \; b | \frac{1}{r_{12}^3} (\mathbf{r}_{12} \times \mathbf{p}_1) \cdot (\mathbf{s}_1 + 2\mathbf{s}_2) | a \; b \rangle \\ &- \langle a \; b | \frac{1}{r_{12}^3} (\mathbf{r}_{12} \times \mathbf{p}_1) \cdot (\mathbf{s}_1 + 2\mathbf{s}_2) | b \; a \rangle \}. \end{aligned} \tag{1.5}$$

where the summations over a and b run over all occupied spin-orbitals and

$$\langle a \; b | \Omega | c \; d \rangle = \int\int \phi_a^*(r_1) \phi_b^*(r_2) \Omega \, \phi_c(r_1) \phi_d(r_2) \mathrm{d}\tau_1 \mathrm{d}\tau_2.$$

For some operator Ω, the functions ϕ are space- and spin-dependent one-electron orbitals and the integrations are over space and spin variables.

The first term in (1.5) is a straightforward one-electron integral to which closed shells of electrons make no contribution;

$$\tfrac{1}{2}Z\alpha^2\sum_a \langle a|\frac{1}{r^3}\mathbf{l}\cdot\mathbf{s}|a\rangle = \tfrac{1}{4}Z\alpha^2 L\left\langle\frac{1}{r^3}\right\rangle.$$

The two-body terms on the other hand are more complicated, corresponding as they do to coulomb and exchange integrals in energy expressions where the operator is the simple electrostatic hamiltonian.

The key to the problem is the rewriting of the operator in tensor form as first done by Blume and Watson, much of whose derivation is reproduced as an appendix to this chapter. There it is shown that

$$\frac{-\alpha^2}{2r_{12}^3}(\mathbf{r}_{12}\times\mathbf{p}_1)\cdot(\mathbf{s}_1+2\mathbf{s}_2) = \mathrm{V}_1+\mathrm{V}_2+\mathrm{V}_3$$

where

$$\mathrm{V}_1=\frac{\alpha^2}{2}\sum_{kq'q\mu}(-1)^k 4\pi\left[\frac{k(k+1)}{2k+1}\right]^{1/2}\begin{pmatrix} k & k & 1\\ q' & q & -\mu\end{pmatrix}$$

$$\times Y_{kq}(\hat{\mathbf{r}}_1)Y_{kq'}(\hat{\mathbf{r}}_2)(\mathbf{s}_1+2\mathbf{s}_2)^{-\mu}\frac{r_<^{k-1}}{r_>^{k+2}}r_2\frac{\partial}{\partial r_1}$$

$$\mathrm{V}_2=\frac{\alpha^2}{2}\sum_{kq'q\mu}(-1)^k 4\pi(2k+1)^{1/2}\left\{-\begin{pmatrix} k-1 & k & 1\\ q' & q & -\mu\end{pmatrix}\right.$$

$$\times\mathrm{T}_{kq}^{(k-1)}(1)Y_{k-1q'}(\hat{\mathbf{r}}_2)\frac{r_1^{k-1}}{r_2^{k+2}}\varepsilon(r_1-r_2)$$

$$\left.+\begin{pmatrix} k+1 & k & 1\\ q' & q & -\mu\end{pmatrix}\mathrm{T}_{kq}^{(k+1)}(1)Y_{k+1q'}(\hat{\mathbf{r}}_2)\frac{r_1^{k-1}}{r_2^{k+2}}\varepsilon(r_2-r_1)\right\}$$

$$\times(\mathbf{s}_1+2\mathbf{s}_2)^{-\mu}$$

$$\mathrm{V}_3=\frac{\alpha^2}{2}\sum_{kq'q\mu}(-1)^k 4\pi(2k+1)^{-1/2}\begin{pmatrix} k & k & 1\\ q' & q & -\mu\end{pmatrix}$$

$$\times\mathrm{T}_{kq}^{(k)}(1)Y_{kq'}(\hat{\mathbf{r}}_2)(\mathbf{s}_1+2\mathbf{s}_2)^{-\mu}$$

$$\times\left[k\frac{r_2^k}{r_1^{k+3}}\varepsilon(r_1-r_2)-(k+1)\frac{r_1^{k-2}}{r_2^{k+1}}\varepsilon(r_2-r_1)\right]$$

and $\epsilon(x) = 1(x > 0); = 0(x < 0)$. Here $\mu = \pm 1, 0$, and the spherical components A^{μ} of a vector $\mathbf{A}$ are defined in terms of the ordinary Cartesian components by

$$A^{\pm 1} = \mp(1/\sqrt{2})(A^{x} \pm iA^{y}), \tag{1.6}$$

$$A^{0} = A^{z}.$$

The $3j$ and $6j$ symbols are defined by Edmonds. The tensor operators $T^{(\lambda)}_{\lambda q}$ are the tensor product of a spherical harmonic with the angular momentum operator $\mathbf{l}$. This particular form of the two-body operator permits the use of the well-developed algebra of Wigner coefficients to be used in the evaluation of the matrix elements.

For example, let us consider the matrix elements

$$\langle a \;\; b|\frac{-\alpha^{2}}{2r_{12}^{3}}(\mathbf{r}_{12}\times\mathbf{p}_{1})\cdot(\mathbf{s}_{1}+2\mathbf{s}_{2})|a \;\; b\rangle.$$

Since these terms are diagonal for both electrons, from the selection rules on spin and orbital magnetic quantum numbers it follows that $q = q' = \mu = 0$ in all the terms of (1.6). This implies that V_1 and V_3 vanish since they contain a factor

$$\begin{pmatrix} k & k & 1 \\ 0 & 0 & 0 \end{pmatrix}$$

which is zero because of the behaviour of the symbol on interchange of columns. The direct term then reduces to

$$\langle a \;\; b|V_2|a \;\; b\rangle$$

Using the expressions derived by Blume and Watson and reproduced in the appendix to this chapter,

$$\langle a \;\; b|V_2|a \;\; b\rangle = -2(m_s^a+2m_s^b)\sum_k(-1)^{k+m^a+m^b}$$

$$\times(2k+1)(2l^a+1)(2l^b+1)[l^a(l^a+1)(2l^a+1)]^{\frac{1}{2}}$$

$$\times\left\{-(2k-1)\begin{pmatrix} l^a & k-1 & l^a \\ 0 & 0 & 0 \end{pmatrix}\begin{pmatrix} l^b & k-1 & l^b \\ 0 & 0 & 0 \end{pmatrix}\right.$$

$$\times\begin{pmatrix} l^a & k & l^a \\ -m^a & 0 & m^a \end{pmatrix}\begin{pmatrix} l^b & k-1 & l^b \\ -m^b & 0 & m^b \end{pmatrix}$$

$$\times\begin{pmatrix} k-1 & k & 1 \\ 0 & 0 & 0 \end{pmatrix}\begin{Bmatrix} l^a & l^a & k \\ 1 & k-1 & l^a \end{Bmatrix} M^{k-1}(ab)$$

$$+(2k+3)\begin{pmatrix} k+1 & k & 1 \\ 0 & 0 & 0 \end{pmatrix}\begin{pmatrix} l^a & k+1 & l^a \\ 0 & 0 & 0 \end{pmatrix}$$

$$\times\begin{pmatrix} l^b & k+1 & l^b \\ 0 & 0 & 0 \end{pmatrix}\begin{pmatrix} l^a & k & l^a \\ -m^a & 0 & m^a \end{pmatrix}$$

$$\times\begin{pmatrix} l^b & k+1 & l^b \\ -m^b & 0 & m^b \end{pmatrix}\begin{Bmatrix} l^a & l^a & k \\ 1 & k+1 & l^a \end{Bmatrix} M^{k-1}(ba)\Bigg\}$$

where the radial integral $M^k(ab)$ is defined by

$$M^k(ab)=(\alpha^2/4)\int_0^\infty\int_0^\infty r_1^2\mathrm{d}r_1 r_2^2\mathrm{d}r_2 f_a^2(r_1)f_b^2(r_2)\frac{r_2^k}{r_1^{k+3}}\,\varepsilon(r_1-r_2).$$

It should be noted that $M^k(ab) \neq M^k(ba)$, except, of course, when f_a and f_b are wave functions for equivalent electrons.

In summing this type of term, only terms with a in an unfilled shell and b in a filled shell or vice versa would appear to contribute but in fact only the former matter since

$$\sum_{m=-l,l}(-1)^m\begin{pmatrix} l & k & l \\ -m & 0 & m \end{pmatrix}=(-1)^l\sqrt{(2l+1)}\,\delta_{k,0}\,.$$

We thus sum over m_a. This requires that $k = 0$ for the terms in the braces, and from the relations of $3j$ and $6j$ symbols,

$$\sum_{m^a}(-1)^{m^a}\begin{pmatrix} l^a & 1 & l^a \\ -m^a & 0 & m^a \end{pmatrix}$$

$$=(-1)^{l^a}[l^a(l^a+1)(2l^a+1)]^{-1/2}\sum_{m^a} m^a$$

$$=(-1)^{l^a}[l^a(l^a+1)(2l^a+1)]^{-1/2}L$$

$$\begin{pmatrix} l^a & 0 & l^a \\ 0 & 0 & 0 \end{pmatrix} = (-1)^{l^a}(2l^a+1)^{-1/2}$$

$$\begin{Bmatrix} l^a & l^b & 1 \\ 1 & 0 & l^a \end{Bmatrix} = -[3(2l^a+1)]^{-1/2}.$$

Using these latter relationships we find that the contribution of the closed shells to the matrix elements is,

$$\sum_{n^b\, l^b} 2(2l^b+1)M^0(n^a\ l^a, n^b\ l^b)L$$

Here $n^a l^a$ are the quantum numbers for the unfilled shell and the summation is over closed shells.

For the exchange terms the summation over k cannot be eliminated.

$$-(\alpha^2/2)\sum_{a\ b}{}' \langle a\ \ b|(1/r_{12}^3)(\mathbf{r}_{12}\times\mathbf{p}_1)\cdot(\mathbf{s}_1+2\mathbf{s}_2)|b\ \ a\rangle$$

$$=\Big(\sum_{(1)}{}' + \sum_{(2)}{}' + \sum_{(3)}{}'\Big)(L/2)$$

where

$$\sum_{(1)}{}' = 6\sum_{b}{}'\sum_{k}(2l^b+1)\left[\frac{2l^a+1}{l^a(l^a+1)}\right]^{1/2}[k(k+1)(2k+1)]^{1/2}$$

$$\times\begin{pmatrix} l^a & k & l^b \\ 0 & 0 & 0 \end{pmatrix}\begin{Bmatrix} l^a & 1 & l^a \\ k & l^b & k \end{Bmatrix} V^{k-1}(ab)$$

$$\sum_{(2)}{}' = -12\sum_{b}{}'\sum_{k}(2l^a+1)(2l^b+1)\frac{(2k-1)(2k+1)}{k+1}$$

$$\times\begin{pmatrix} l^a & k-1 & l^b \\ 0 & 0 & 0 \end{pmatrix}^2\begin{Bmatrix} l^b & l^a & k \\ 1 & k-1 & l^a \end{Bmatrix} N^{k-1}(ab)$$

$$\sum_{(3)}{}' = 6 \sum_b{}' \sum_k (2l^b+1) \frac{(2k+1)}{[k(k+1)(2k+1)]^{1/2}} \left[\frac{2l^a+1}{l^a(l^a+1)} \right]^{1/2}$$

$$\times \begin{pmatrix} l^a & k & l^b \\ 0 & 0 & 0 \end{pmatrix} \begin{Bmatrix} l^a & 1 & l^a \\ k & l^b & k \end{Bmatrix}$$

$$\times [l^b(l^b+1) - l^a(l^a+1)][-kN^k(ab) + (k+1)N^{k-2}(ab)].$$

Here a stands for $n^a l^a$, the quantum numbers of the outer electrons, and the summation over b is over all closed shells. The dash indicates that, as for the direct terms, we have included only the contributions to the summation which arise when one electron is in the core and the other is in the unfilled shell. The exchange integrals N^k and V^k are defined by

$$V^k(ab) = V^k(n^a l^a, n^b l^b)$$

$$= (\alpha^2/4) \int_0^\infty \int_0^\infty r_1^2 \mathrm{d}r_1 r_2^2 \mathrm{d}r_2 f_{n^a l^a}(r_1) f_{n^b l^b}(r_2)$$

$$\times \frac{r_<^k}{r_>^{k+3}} \left(r_2 \frac{\partial}{\partial r_1} - r_1 \frac{\partial}{\partial r_2} \right) f_{n^a l^a}(r_2) f_{n^b l^b}(r_1)$$

$$N^k(ab) = N^k(n^a l^a, n^b l^b)$$

$$= (\alpha^2/4) \int_0^\infty \int_0^\infty r_1^2 \mathrm{d}r_1 r_2^2 \mathrm{d}r_2 f_{n^a l^a}(r_1) f_{n^b l^b}(r_2)$$

$$\times f_{n^a l^a}(r_2) f_{n^b l^b}(r_1) \frac{r_2^k}{r_1^{k+3}} \varepsilon(r_1 - r_2).$$

They have the properties

$$N^k(ab) = N^k(ba)$$

$$V^k(ab) = -V^k(ba)$$

$$N^k(aa) = M^k(aa)$$

1.4.3. *Results*

A summary is given in Table 1.2 of the salient points of the results found by Blume and Watson using the above formulae and the non-relativistic analytic Hartree–Fock radial wavefunctions for atoms. The experimental results are simply obtained from the Condon and Shortley formula

$$E_J - E_{J-1} = \lambda J,$$

and the variations of λ give some idea of the breakdown of Russell–Saunders coupling and the importance of spin–spin splitting.

Table 1.2

Observed and calculated spin–orbit coupling in atoms

Atom	Configuration	Observed	Calculated
B	$(2p)^1$	10.7	9.74
C	$(2p)^2$	13 to 16	13.4
N^+	$(2p)^2$	41 to 49	41.8
O^{2+}	$(2p)^2$	97 to 113	99.2
F^{3+}	$(2p)^2$	194 to 225	200
O	$(2p)^4$	−68 to −79	−79.6
F^+	$(2p)^4$	−149 to −170	−168
F	$(2p)^5$	−269	−265
Sc^{2+}	$(3d)^1$	79	85.7
Ti^{2+}	$(3d)^2$	59-61	61
V^{2+}	$(3d)^3$	56	57.0
Cr^{2+}	$(3d)^4$	54-61	59.1
Fe^{2+}	$(3d)^6$	−94 to −109	−114
Co^{2+}	$(3d)^7$	−166 to −186	−189
Ni^{2+}	$(3d)^8$	−303 to −340	−343
Cu^{2+}	$(3d)^9$	−829	−830
Al	$(3p)^1$	74.6	60.5
Si	$(3p)^2$	73 to 77	64.0
S	$(3p)^4$	−176 to −198	−184
Cl	$(3p)^5$	−587	−545
Ga	$(4p)^1$	551	460
Ge	$(4p)^2$	440	399
Se	$(4p)^4$	−934	−825
Br	$(4p)^5$	−2456	−2194

1.5. Conclusions

In general the agreement between calculated and observed spin-orbit coupling is very good. Poorer results are however obtained with 3p and 4p electrons. One possible reason for this is error in $\langle r^{-3} \rangle$ arising from inaccuracies in the wavefunction at the nucleus. Another possibility is 'core polarization'; the inner p shells do not give zero contributions since the interaction of an outer electron of the same symmetry may produce differences in the otherwise identical radial functions of the closed shells.

The quality of the agreement also provides the impetus to extend this type of work into the realm of diatomic molecules where not only is there an abundance of problems, but also the *ab initio* molecular wavefunctions are expressed as linear combinations of atomic orbitals and hence much of the analysis presented in this chapter may be transferred.

Further reading

1. Condon, E. U. and Shortley, G. H., *The theory of atomic spectra*. Cambridge University Press (1963).
2. Bethe, H. A. and Salpeter, E. E., *Quantum mechanics of one and two electron atoms*. Academic Press, New York (1957).
3. Edmonds, A. R., *Angular momentum in quantum mechanics*. Princeton University Press (1957).
4. Brink, D. M. and Satchler, G. R., *Angular momentum*. Clarendon Press, Oxford (1968).
5. Judd, B. R., *Operator techniques in atomic spectroscopy*. McGraw-Hill, New York (1963).
6. Blume, M. and Watson, R. E., Theory of spin-orbit coupling in atoms. I. Derivation of the spin-orbit coupling constant. *Proc. R. Soc. (Lond.)* **A 270**, 127 (1962).
7. Blume, M. and Watson, R. E., Theory of spin-orbit coupling in atoms. II. Comparison of theory with experiment. *Proc. R. Soc. (Lond.)* **A 271**, 565 (1963).
8. Horie, H., Spin-spin and spin-other-orbit interactions. *Prog. theoret. Phys.* **10**, 296 (1953).

APPENDIX 1.A

DERIVATION OF AN EXPRESSION FOR THE SPIN-OTHER ORBIT OPERATOR IN TERMS OF SPHERICAL HARMONICS

To express the operator

$$V=(\alpha^2/2)\nabla_i(1/r_{ij})\times\mathbf{p}_i\cdot(\mathbf{s}_i+2\mathbf{s}_j)$$

in tensor operator forms, we write

$$(1/r_{ij})=\sum_k \frac{r_<^k}{r_>^{k+3}}(C_i^k\cdot C_j^k)$$

using the gradient operator in the form

$$\nabla f(r)C^l(r)=\sum_\lambda (2\lambda+1)\begin{pmatrix} l & 1 & \lambda \\ 0 & 0 & 0 \end{pmatrix} C^\lambda(r)\mathrm{D}^{\lambda,l}f(r)$$

where

$$\mathrm{D}^{\lambda,l}=\frac{\partial}{\partial r}+E(\lambda,l)\frac{1}{r}$$

and

$$E^{\lambda,l}=\sqrt{6}[l(l+1)(2l+1)]^{\frac{1}{2}}\begin{pmatrix} l & 1 & \lambda \\ 1 & l & 1 \end{pmatrix}$$

from the 3j symbol $\lambda=l\pm 1$

$$E(l+1,l)=-l \qquad E(l-1,l)=l+1$$

hence

$$\nabla_i(1/r_{ij})=\frac{1}{\sqrt{3}}\sum_k(-1)^k\left[\frac{r_i^{k-1}}{r_j^{k+1}}[k(2k-1)(2k+1)]^{\frac{1}{2}}\{C_i^{k-1}C_j^k\}^1\right.$$

$$\left.+\frac{r_j^k}{r_i^{k+2}}[(k+1)(2k+1)(2k+3)]^{\frac{1}{2}}\{C_i^{k+1}C_j^k\}^1\right]. \quad (1.A.1)$$

The first term is to be used if $r_j > r_i$ and the second if $r_i > r_j$.

Now

$$\nabla_i(1/r_{ij})\times \mathbf{p}_i = -i\sqrt{2}\{\nabla_i(1/r_{ij})\mathbf{p}_i\}^1$$

so that

$$\frac{1}{i\sqrt{2}}\nabla_i(1/r_{ij})\times \mathbf{p}_i = \{\{C_i^{k-1}C_j^k\}^1\mathbf{p}_i\}^1 \qquad (1.A.2)$$

, etc.

recoupling

$$\frac{1}{i\sqrt{2}}\nabla_i(1/r_{ij})\times \mathbf{p}_i = \sum_k \sqrt{(3(2\lambda+1))}\begin{Bmatrix} 1 & 1 & 1 \\ k & k-1 & \lambda \end{Bmatrix}$$
$$\times\{C_j^k(C_i^{k-1}\mathbf{p}_i)^\lambda\}^1 \qquad (1.A.3)$$

using the expression for the momentum

$$\mathbf{p} = (1/r)(\hat{\mathbf{r}}\times\mathbf{l}) - \hat{\mathbf{r}}(\hat{\mathbf{r}}\cdot\mathbf{p})$$
$$= (1/r)(\hat{\mathbf{r}}\times\mathbf{l}) - i\hat{\mathbf{r}}\frac{\partial}{\partial r}$$
$$= (i\sqrt{2}/r)(C^1 l)^1 - iC^1\frac{\partial}{\partial r}$$

hence

$$\{C_i^{k-1}p_i\}^\lambda = \sqrt{2}(i/r)\{C_i^{k-1}\{C_i^1 l\}^1\}^\lambda - i\{C_i^{k-1}C_i^1\}\frac{\partial}{\partial r}.$$

Consider the second term on the right (part B)

$$\mathrm{B} = i\{C_i^{k-1}C_i^1\}^\lambda\frac{\partial}{\partial r} = i(-1)^\lambda\sqrt{(2\lambda+1)}\begin{pmatrix} k-1 & \lambda & 1 \\ 0 & 0 & 0 \end{pmatrix} C_i^\lambda.$$

Recoupling the first term on the right

$$\mathrm{A} = \sqrt{2}(i/r)\{C_i^{k-1}\{C_i^1 l\}^1\}^\lambda$$

$$= \sqrt{2}(i/r)\sum_\lambda (-1)^{k-1+\lambda}\sqrt{(3(2\lambda'+1))}\begin{Bmatrix} k-1 & \lambda & 1 \\ 1 & 1 & \lambda' \end{Bmatrix}$$

and

$$\times\{\{C_i^{k-1}C_i^1\}^\lambda l\}^\lambda$$

$$\{C_i^{k-1}C_i^1\}^{\lambda'} = (-1)^{\lambda'}\sqrt{(2\lambda'+1)}\begin{pmatrix} k-1 & \lambda' & 1 \\ 0 & 0 & 0 \end{pmatrix} C_i^{\lambda'}.$$

substituting these expressions into (1.A.3) we obtain two parts.

Firstly,

$$B' = i \sum_{\lambda} (-1)^{\lambda+1} 3(2\lambda+1) \begin{pmatrix} k-1 & \lambda & 1 \\ 0 & 0 & 0 \end{pmatrix}$$

$$\times \begin{Bmatrix} 1 & 1 & 1 \\ k & k-1 & \lambda \end{Bmatrix} \{C_i^{\lambda} C_j^{k}\}^1 \frac{\partial}{\partial r}.$$

From the $3j$ symbol $\lambda = k$ or $k+2$: from the $6j$ symbol $\lambda = k$ or $k+1$.

Hence $\lambda = k$

and

$$B' = -i \left[\frac{k+1}{2(2k-1)} \right]^{1/2} \{C_i^k C_j^k\}^1 \frac{\partial}{\partial r}.$$

Secondly

$$A' = \sqrt{2}(i/r) \sum_{\lambda\lambda'} (-1)^{k+\lambda+\lambda'} 3(2\lambda'+1)\sqrt{(2\lambda+1)}$$

$$\times \begin{pmatrix} k-1 & \lambda' & 1 \\ 0 & 0 & 0 \end{pmatrix} \begin{pmatrix} 1 & 1 & 1 \\ k & k-1 & \lambda \end{pmatrix}$$

$$\times \begin{pmatrix} k-1 & \lambda & 1 \\ 1 & 1 & \lambda' \end{pmatrix} \{\{C_i^{\lambda'} l\} C_j^k\}^1.$$

From the $6j$ symbols $\lambda = k$ or $k-1$: when $\lambda = k$, $\lambda' = k$; when $\lambda = k-1$, $\lambda' = k$ or $k-2$.

(i) for $\lambda = k-1$

$$A'' = -\sqrt{2}(i/r) \left(\frac{k-1}{2(2k+1)} \left[\frac{2k+1}{k} \right]^{1/2} \{\{C_i^k l\}^{k-1} C_j^k\}^1 \right.$$

$$\left. + \frac{[(2k-3)(2k-1)]^{1/2}}{2(2k-1)} \{\{C_i^{k-2} l\}^{k-1} C_j^k\}^1 \right)$$

using the relation

$$\{C^k l\}^{k-1} = \frac{[(k-1)(2k-3)]^{1/2}}{k(2k+1)} \{C^{k-2} l\}^{k-1}$$

$$A'' = -\frac{i\sqrt{(2k+1)}}{2kr} \{\{C_i^k l\}^{k-1} C_j^k\}^1.$$

(ii) for $\lambda = k$

$$A''' = \frac{i(k+1)}{r} \left[\frac{k}{2(2k+1)}\right]^{1/2} \{\{C_i^k l\}^k C_j^k\}^1.$$

Substituting A'', A''', and B' into (1.A.2) we find

$$\sqrt{3}\nabla_i(1/r_{ij}) \times \mathbf{p}_i = \sum_k (-1)^k [(1/r_i)(k+1)\sqrt{(2k+1)} \{\{C_i^k l\}^k C_j^k\}^1$$

$$-(1/r_i)(2k+1)\sqrt{(2k-1)} \{\{C_i^k l\}^{k-1} C_j^k\}^1$$

$$-[(k+1)k(2k+1)]^{\frac{1}{2}} \{C_i^k C_j^k\}^1 \frac{\partial}{\partial r}] \frac{r_i^{k-1}}{r_j^{k+1}} \qquad \text{for } r_i < r_j$$

and a similar expression for $r_i > r_j$.

These expressions allow the calculation of reduced matrix elements. However, since we are interested in matrix elements between atomic basis orbitals, the full expressions are needed. Now

$$\mathbf{a} \cdot \mathbf{b} = \sum_\mu (-1)^\mu a^\mu b^{-\mu}$$

and

$$X_Q^k = \sum_{qq'} \sqrt{(2k+1)} \begin{pmatrix} k_2 & k_1 & k \\ q' & q & -Q \end{pmatrix} T_q^{k_1} U_{q'}^{k_2}$$

This latter expression gives an operator X in terms of its two separable parts T and U.

It is convenient to use the same notation as Blume and Watson, because the subroutine used in the programs generates the spherical harmonics in this manner. Writing

$$Y_{kq} = \left[\frac{(2k+1)}{4\pi}\right]^{1/2} C_q^k$$

and

$$T_{kq}^{k-1} = \left[\frac{(2k-1)}{4\pi}\right]^{1/2} \{C^{k-1} l\}_q^k$$

We arrive at the following expressions for the spin–other-orbit operator,

$$V = V_1 + V_2 + V_3,$$

where

$$V_1 = \frac{\alpha^2}{2} \sum_{kq'\mu} (-1)^k 4\pi \left[\frac{k(k+1)}{2k+1}\right]^{1/2} \begin{pmatrix} k & k & 1 \\ q' & q & -\mu \end{pmatrix}$$

$$\times Y_{kq}(\hat{\mathbf{r}}_1) Y_{kq'}(\hat{\mathbf{r}}_2)(\mathbf{s}_1 + 2\mathbf{s}_2)^{-\mu} \frac{r_<^{k-1}}{r_>^{k+2}} r_2 \frac{\partial}{\partial r_1}$$

$$V_2 = \frac{\alpha^2}{2} \sum_{kq'\mu} (-1)^k 4\pi (2k+1)^{1/2} \Big\{ -\begin{pmatrix} k-1 & k & 1 \\ q' & q & -\mu \end{pmatrix}$$

$$\times T_{kq}^{(k-1)}(1) Y_{k-1q}(\hat{\mathbf{r}}_2) \frac{r_1^{k-1}}{r_2^{k+2}} \varepsilon(r_1 - r_2)$$

$$+ \begin{pmatrix} k+1 & k & 1 \\ q' & q & -\mu \end{pmatrix} T_{kq}^{(k+1)}(1) Y_{k+1q'}(\hat{\mathbf{r}}_2) \frac{r_1^{k-1}}{r_2^{k+2}} \varepsilon(r_2 - r_1) \Big\}$$

$$\times (\mathbf{s}_1 + 2\mathbf{s}_2)^{-\mu}$$

$$V_3 = \frac{\alpha^2}{2} \sum_{kq'\mu} (-1)^k 4\pi (2k+1)^{-1/2} \begin{pmatrix} k & k & 1 \\ q' & q & -\mu \end{pmatrix}$$

$$\times T_{kq}^{(k)}(1) Y_{kq'}(\hat{\mathbf{r}}_2)(\mathbf{s}_1 + 2\mathbf{s}_2)^{-\mu}$$

$$\times \left[k \frac{r_2^k}{r_1^{k+3}} \varepsilon(r_1 - r_2) - (k+1) \frac{r_1^{k-2}}{r_2^{k+1}} \varepsilon(r_2 - r_1)\right]$$

In V_2 and V_3 the first term refers to $r_1 > r_2$ and the second to $r_2 > r_1$.

Matrix elements of $T^{\lambda}_{\lambda'\sigma'}$ are given by

$$\langle l' \; m' | T^{\lambda}_{\lambda'\sigma'} | l \; m \rangle = (-1)^{\lambda+\lambda'+m'}(2l+1)$$

$$\times \left[\frac{-(2\lambda+1)(2\lambda'+1)l(l+1)(2l+1)}{4\pi} \right]^{1/2} \begin{pmatrix} l' & \lambda & l \\ 0 & 0 & 0 \end{pmatrix}$$

$$\times \begin{pmatrix} l' & \lambda' & l \\ -m' & \sigma' & m \end{pmatrix} \begin{Bmatrix} l' & l & \lambda' \\ 1 & \lambda & l \end{Bmatrix}$$

and of Y_{lm} by

$$\langle l_1 \; m_1 | Y_{lm} | l_2 \; m_2 \rangle = (-1)^{m_1} \left[\frac{(2l_1+1)(2l+1)(2l_2+1)}{4\pi} \right]^{1/2}$$

$$\times \begin{pmatrix} l_1 & l & l_2 \\ -m_1 & m & m_2 \end{pmatrix} \begin{pmatrix} l_1 & l & l_2 \\ 0 & 0 & 0 \end{pmatrix}.$$

These expressions agree with those derived by Horie. His operator U^{Kk} is related to the ones employed here by

$$U^{k-1,k} = \frac{1}{(2k-1)} \left[\frac{3}{(2k+1)} \right]^{1/2} \{C^{k-1} l\}^{k}$$

$$= \left[\frac{3}{4\pi(2k+1)(2k-1)} \right]^{1/2} T^{k-1}_{k}.$$

It should be noted that there is a misprint in the paper by Thorhallson *et al.* (*J. chem. Phys.* **48**, 2925 (1968)). The correct matrix elements are:

$$\langle l \| U^{k-1,k} \| l \rangle = (-1)^{(k-1)/2}(2l+1)[(k+1)(2k-1)(2k+1)/3]^{1/2}$$

$$\times \left[\frac{(k-1)!(k+1)!(2l-k)!}{(2l+k+1)!} \right]^{1/2}$$

$$\times \frac{[(l+(k+1)/2]!}{[(k-1)/2]![(k+1)/2]![l-(k+1)/2]!}$$

2

DIAGONAL SPIN–ORBIT COUPLING EFFECTS IN MOLECULES

2.1. Introduction

THE spin–orbit coupling effect manifests itself in linear molecules in much the same way as it does in atoms. Fig. 2.1 shows the multiplet splittings in a $^3\Pi$ state and should be compared to Fig. 1.3 which

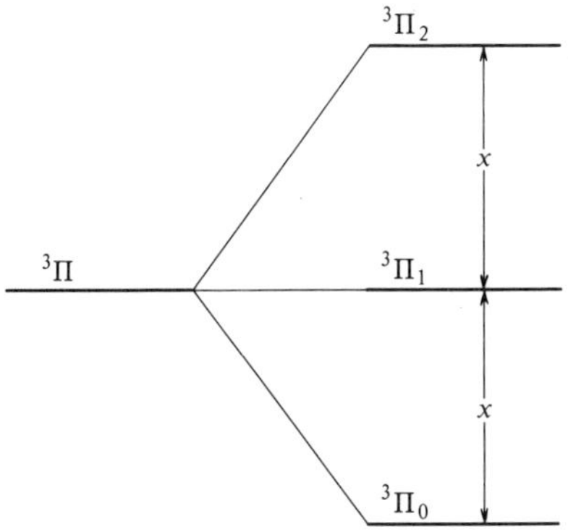

Fig. 2.1. Multiplet splitting in a $^3\Pi$ state.

shows the multiplet splittings in a ^3P state of an atom. The Landé interval rule which applies for atoms is not applicable to molecules: the axial symmetry causes the sublevels of $^3\Pi$ to be equidistant.

It is usual to define a spin–orbit coupling constant A by equating the hyperfine spacings to $2A\Lambda\Sigma$ where Λ is the component of the total electronic orbital angular momentum and Σ is the component of the total electronic spin angular momentum along the internuclear axis. As in atoms, A is positive for regular states (less than half-filled shells) and negative for inverted states (electronic shells more than half-filled). The sign of the spin–orbit coupling constant thus provides a convenient and powerful way of learning whether a state is regular or inverted. Since spin–orbit coupling constants are usually very different for different electronic states, knowledge of A is important in the identification of states.

In this chapter, we shall deal with situations involving diagonal matrix elements of the spin–orbit coupling operator. Much of the analysis carries over directly to the calculation of off-diagonal matrix elements; these are important in phenomena such as Λ-doubling which are considered in the next chapter.

As a first approximation, molecular spin–orbit coupling constants may be obtained from atomic spin–orbit coupling constants and Mulliken population analyses. This is reasonable in cases where the contribution from one atom swamps that from the other but the idea of charge localized on atoms must be treated with some suspicion. We shall describe *ab initio* methods of calculating the spin–orbit coupling constant A.

2.2. The coupling of angular momenta and molecular wavefunctions

The various angular momenta present in a molecule include the total electron spin **S**, the total orbital angular momentum **L**, and the nuclear framework momentum **O**. Hund has defined four coupling schemes for diatomic molecules: few molecules actually obey one of these but many molecules closely resemble one of the four cases rather more than the others. For molecules such as CH, OH, and SiH, Hund's cases (a) and (b) are the most appropriate.

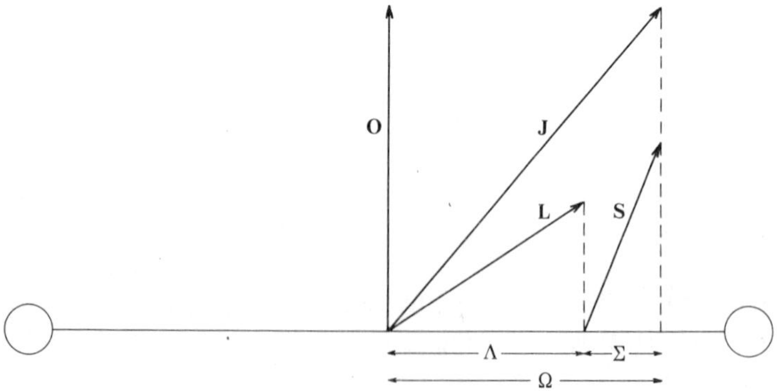

Fig. 2.2. Hund's coupling case (a).

In Hund's case (a), the spin–orbit coupling constant is large compared with the rotational constant, B. In other words, the energy arising from spin–orbit interaction is much greater than the energy due to rotation of the nuclear framework. The axial electric field is large and couples **L** (and thus **S**) to the molecular axis. The projections on to the axis, Λ and Σ, give a resultant Ω which couples with **O** to produce the total angular momentum **J**.

In Hund's case (b), rotational energy is large compared to the spin–orbit coupling energy. The dividing line between cases (a) and (b) is when A is twice B. **L** is still coupled electrostatically to the axis, but the interaction between **L** and **S** is no longer sufficient to couple **S** to the axis. This occurs when there is no electronic orbital angular momentum or when the spin–orbit coupling constant is small or when the rotation quantum number is large. The electronic orbital momentum couples with the nuclear framework momentum to form a resultant **N** which then couples with the spin momentum to form **J**.

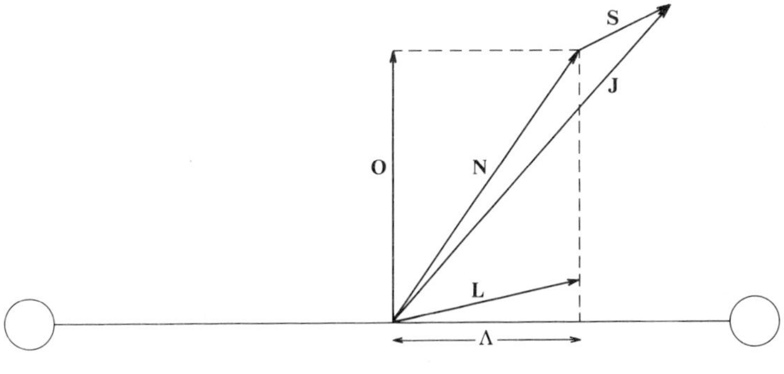

Fig. 2.3. Hund's coupling case (b).

In Hund's case (c), the spin–orbit interaction is strong. This occurs, for example, when heavy atoms are present. **L** and **S** are coupled to form a resultant $\mathbf{J}_a$. The projection of this total electronic momentum onto the axis, Ω, then couples with **O** to form **J**.

In Hund's case (d), the electronic orbital angular momentum is no longer coupled to the molecular axis. Spin–orbit coupling is small and so **L** couples with **O** to form a resultant **N** which then couples with **S** to form **J**. This scheme is important for Rydberg states.

Figs. 2.2–2.5 illustrate schematically the order in which the various angular momenta are coupled in the different coupling cases.

The molecular wavefunctions are built from atomic wavefunctions by making a linear combination which satisfies the proper symmetry

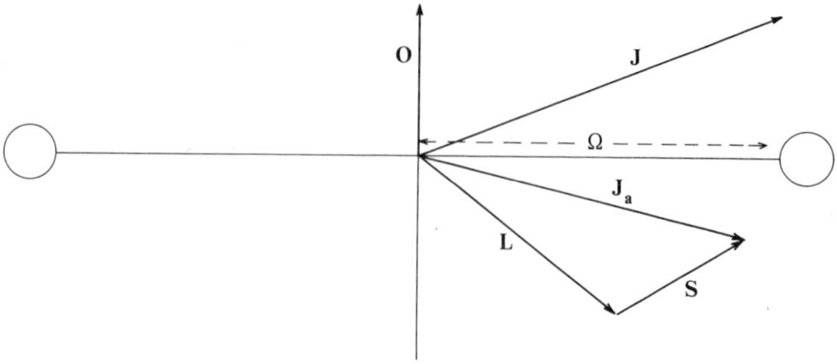

Fig. 2.4. Hund's coupling case (c).

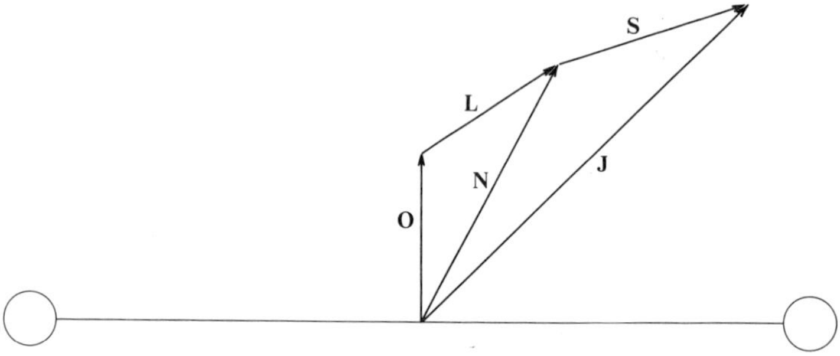

Fig. 2.5. Hund's coupling case (d).

requirements. These are called LCAO–MO wavefunctions. If in addition they satisfy a set of self-consistent field (SCF) equations – as they usually do – they are called LCAO–MO–SCF wavefunctions or just SCF wavefunctions. For a closed shell molecule, the SCF wavefunction ψ can be represented by a single determinant,

$$\Psi(\mathbf{r}_1,\mathbf{r}_2,\ldots,\mathbf{r}_n) = \frac{1}{\sqrt{n!}} \begin{vmatrix} \phi'_1(\mathbf{r}_1) & \phi'_1(\mathbf{r}_2) & \ldots & \phi'_1(\mathbf{r}_n) \\ & & & \\ \phi'_n(\mathbf{r}_1) & \phi'_n(\mathbf{r}_2) & \ldots & \phi'_n(\mathbf{r}_n) \end{vmatrix} \tag{2.1}$$

made up of single particle molecular orbitals ϕ' which in turn are made up of a spin function Θ and a spatial part ϕ made up of atomic orbitals χ

$$\phi_i = \sum_k c_{ik}\chi_k .$$

We have assumed Slater-type atomic orbitals (STO) in what follows:

$$\chi_{nlm} = k_n Y_{lm}(\theta,\phi)\exp(-\zeta r)r^{n-1} \tag{2.2}$$

where $Y_{lm}(\theta, \phi)$ is a spherical harmonic; n, l, and m are integer quantum numbers ($n = 1, 2, 3, \ldots ; l = 0, 1, 2, \ldots, n-1; m = -l, -l+1, \ldots, l$); k_n is chosen so that $\langle\chi_{nlm}|\chi_{nlm}\rangle = 1$ and ζ is a parameter. A more elaborate type of molecular wavefunction which takes into account electron correlation can be constructed by taking a linear combination of determinants like (2.1)

$$\Psi_\mu = \sum_k D_{\mu k}\Psi_k .$$

This is an SCF–CI (configuration interaction) wavefunction. The determinants Ψ_k are most commonly constructed by exciting one or more of the electrons into the virtual states, obtained as a by-product of the SCF calculation; the advantages being that (i) this can be done with little extra labour and (ii) the wavefunctions so constructed are automatically orthogonal to each other. This greatly simplifies the evaluation of matrix elements between such wavefunctions. The disadvantage is that such excited state wavefunctions are more akin to those of the negative ion than to those of the original system, resulting in rather slow and often non-uniform convergence. Determinants, each made up of a different basic configuration chosen to represent the excited state more closely than virtual orbitals, are sometimes used. This type of function is known as the multi-configuration, self-consistent (MC–SCF) wavefunction. While a much improved representation of the excited state is achieved and the convergence improves, the MC–SCF wavefunctions are much more difficult to compute and the evaluation of matrix elements with them becomes very laborious. Only SCF wavefunctions with or without CI will be assumed in our calculations.

2.3. Spin–orbit coupling operators

Transformation of the relativistic many-electron Breit equation to a zero-order, non-relativistic equation plus relativistic corrections produces two terms which contribute to spin–orbit effects. These are jointly written H_{SO}

$$H_{SO} = \sum_i h_i + \sum_{i \neq j} H_{ij}$$

where

$$h_i = (\alpha^2/2)\sum_N (Z_N/r_{iN}^3)(\mathbf{r}_{iN}\times\mathbf{p}_i)\cdot\mathbf{s}_i \tag{2.3}$$

and

$$H_{ij} = -(\alpha^2/2)(1/r_{ij}^3)(\mathbf{r}_{ij}\times\mathbf{p}_i)\cdot(\mathbf{s}_i+2\mathbf{s}_j). \tag{2.4}$$

In the above expressions, α is the fine structure constant, Z_N is the charge on nucleus N, i and j label electrons, and **r**, s, and **p** are one-electron position vector, spin and linear momentum operators. The manipulation of angular integrals is simplified by rewriting the operators in spherical tensor notation,

$$h_i = (\alpha^2/2)\sum_N \sum_\mu (-1)^\mu (Z_N/r_{iN}^3)(\mathbf{r}_{iN}\times\mathbf{p}_i)^\mu(\mathbf{s}_i)^{-\mu}$$

and

$$H_{ij} = -(\alpha^2/2)\sum_\mu (-1)^\mu (1/r_{ij}^3)(\mathbf{r}_{ij}\times\mathbf{p}_i)^\mu(\mathbf{s}_i+2\mathbf{s}_j)^{-\mu}.$$

The similarity to the atomic case should be obvious and we shall find much of the discussion in Chapter 1 very useful.

Veseth has investigated the relation in a diatomic molecule between the simple phenomenological operator $A\mathbf{L}\cdot\mathbf{S}$ and the Pauli–Breit operator and has found that both operators lead formally to the same results.

2.4. Matrix elements: one- and two-centre integrals

The matrix element of an n-body operator can only be non-zero if the two states differ by no more than n single-particle wavefunctions. This is because an n-body operator can at most change n single-particle wavefunctions. Thus, in evaluating the matrix elements of

h_i and H_{ij} (eqns (2.3) and (2.4)), we need only consider states which differ from each other by at most one and two spin-orbitals. Slater and Löwdin have prescribed rules which facilitate the evaluation of one- and two-electron matrix elements. The matrix elements of h_i and H_{ij} ultimately reduce to combinations of matrix elements of the type $\langle \phi_i'(1) | h_1 | \phi_j'(1) \rangle$, involving the spin orbitals of only one electron and $\langle \phi_i'(1)\phi_j'(2) | H_{12} | \phi_k'(1)\phi_l'(2) \rangle$ involving the spin orbitals of only two electrons. The spin part and the spatial parts separate out and the spin integration is readily performed analytically. Since the molecular orbitals are linear combinations of atomic orbitals, the spatial integration reduces to integrals evaluated over atomic orbital basis functions.

The one-electron operator h_1' is the simpler of the two operators making up H_{SO} and frequently makes larger contributions to the energy than does the two-electron operator H_{12}'. The prime serves to remind us that we are dealing with the spatial part only. We shall also drop α^2 from explicit formulae but its presence must be understood. Denoting the atom (*A* or *B*) by a subscript, three kinds of integrals involving h_1' are seen to arise:

$\langle \chi_A(1) | h_{A1}'^{\mu} | \chi_{A'}(1) \rangle_O$ the one-centre integral,

$\langle \chi_A(1) | h_{B1}'^{\mu} | \chi_{A'}(1) \rangle_H$ the Coulomb integral, and

$\langle \chi_A(1) | h_{A1}'^{\mu} | \chi_B(1) \rangle_C$ the hybrid integral, where

$$h_{A1}'^{\mu} = (Z_A/2r_{A1}^3) l_{A1}^{\mu}$$

$$\mathbf{l} = \mathbf{r} \times \mathbf{p}$$

The one-centre integral is readily evaluated as

$$\langle \chi_A(1) | h_{A1}'^{\mu} | \chi_{A'}(1) \rangle = \tfrac{1}{2} k_{nA} k_{nA'} Z_A \frac{(n_A + n_{A'} - 3)!}{(\zeta_A + \zeta_{A'})^{n_A + n_{A'} - 2}}$$

$$\times \langle l_A \| l \| l_{A'} \rangle \langle l_A m_A | l_{A'} 1 m_{A'} \mu \rangle$$

using the Wigner–Eckert theorem. This is not the expression that we shall use, since it is desirable to make the one- and the two-centre integral programs compatible. We follow the works of Matcha and co-workers and rewrite

$$\mathrm{h}'^{\mu}_{A1} = (2\pi)^{\frac{1}{2}} Z_A(-1)^{\mu} \sum_{\beta\beta'} \begin{pmatrix} 1 & 1 & 1 \\ \beta & -\beta' & \mu \end{pmatrix} Y_{1-\beta}(\hat{\mathbf{r}}_{A1})\ r_{A1}^{-2}\ \nabla^{\beta'} \tag{2.5}$$

using the algebra of spherical tensor operators. Here ( ) stands for the 3*j* symbol. Then

$$\langle \chi_A | \mathrm{h}'^{\mu}_{A} | \chi_{A'} \rangle = (2\pi)^{\frac{1}{2}} Z_A(-1)^{\mu} \sum_{\beta\beta'} \left\{ \begin{pmatrix} 1 & 1 & 1 \\ \beta & -\beta' & \mu \end{pmatrix} \times \int \chi^{*}_{A} Y_{1-\beta}(\hat{\mathbf{r}}_A)\ r_A^{-2}\ \nabla^{\beta'} \chi_{A'} \mathrm{d}V \right\} \tag{2.6}$$

To proceed further it becomes necessary to specify the form of the basis functions. For the STO defined by (2.2)

$$\nabla^{\beta'} \chi_{n'l'm'} = \sum_{\bar{n}\bar{l}} c^{\beta'}[n'l'm'|\bar{n}\bar{l}\ m'+\beta']\chi_{\bar{n}\bar{l}m'+\beta'} \tag{2.7}$$

where

$$c^{\beta'}[n'l'm'|\bar{n}\bar{l}\ m'+\beta'] = (-1)^{m'+\beta'}[(2l'+1)(2\bar{l}+1)]^{\frac{1}{2}} \times \begin{pmatrix} l' & 1 & \bar{l} \\ m' & \beta' & -m'-\beta' \end{pmatrix} \begin{pmatrix} l' & 1 & \bar{l} \\ 0 & 0 & 0 \end{pmatrix} \times \begin{cases} \{2\zeta'/[2n'(2n'-1)]^{\frac{1}{2}}\}\{n'-1+C(\bar{l},l')\} & \text{for } \bar{n}=n'-1 \\ -\zeta' & \text{for } \bar{n}=n \\ 0 & \text{otherwise} \end{cases} \tag{2.8a}$$

and

$$C(\bar{l},l') = \begin{cases} -l' & \text{for} \quad \bar{l}=l'+1 \\ l'+1 & \text{for} \quad \bar{l}=l'-1 \\ 0 & \text{otherwise} \end{cases} \tag{2.8b}$$

Using (2.8) and the identity

$$\langle l\ m|Y_{1-\beta}|l'\ m'+\beta'\rangle = (-1)^m[3(2l+1)(2l'+1)/(4\pi)]^{\frac{1}{2}} \times \begin{pmatrix} l & 1 & l' \\ m & \beta & -m'-\beta' \end{pmatrix} \begin{pmatrix} l & 1 & l' \\ 0 & 0 & 0 \end{pmatrix}$$

we obtain

$$\langle \chi_A|h'^{\mu}_{A1}|\chi_{A'}\rangle_0 = \tfrac{1}{2}[6(2l+1)]^{\frac{1}{2}} Z_A(-1)^{m+\mu}\delta(m',m-\mu)k_n \times \sum_{\beta} \begin{pmatrix} 1 & 1 & 1 \\ \beta & -\beta-\mu & \mu \end{pmatrix} \sum_{\bar{n}\bar{l}} K^{\beta+\mu}_{m\beta}\,[n'l'm'|\bar{n}\bar{l}\ m'+\beta+\mu] \times (n+\bar{n}-2)!/(\zeta+\zeta')^{n+\bar{n}-1} \tag{2.9}$$

where

$$K^{\beta'}_{m\beta}\,[n'l'm'|\bar{n}\bar{l}\ m'+\beta'] = (2\bar{l}+1)^{\frac{1}{2}}\, k_{\bar{n}} c^{\beta'}[n'l'm'|\bar{n}\bar{l}\ m'+\beta'] \times \begin{pmatrix} l & 1 & \bar{l} \\ m & \beta & -m'-\beta' \end{pmatrix} \begin{pmatrix} l & 1 & \bar{l} \\ 0 & 0 & 0 \end{pmatrix} \tag{2.10}$$

The sums over $\bar{n}$ and $\bar{l}$ range over the values $\bar{n} = n'-1, n'$ and $\bar{l} = l'\pm 1$, provided $\bar{n} \geqslant \bar{l} \geqslant |m'+\beta'|$.

The coulomb integral is best dealt with by expanding the operator $Y_{1-\beta}(\hat{\mathbf{r}}_B)/r_B^2$ about the centre A using the identities

$$Y_{lm}(\hat{\mathbf{r}}_B)/r_B^2 = \sum_{\lambda=|m|,\infty} \left\{ \frac{[(2l+1)/(2\lambda+1)]}{[(\lambda-m)!(\lambda+m)!(l-m)!(l+m)!]} \right\}^{1/2}$$

$$\times\ (\lambda+l)!r_A^{\lambda}R^{-\lambda-l-1}Y_{\lambda m}(\hat{\mathbf{r}}_A) \qquad \text{for } r_A \leqq R$$

(2.11a)

$$Y_{lm}(\hat{\mathbf{r}}_B)/r_B^2 = (-1)^{l+m} \sum_{\lambda=l,\infty} \left\{ \frac{(\lambda-m)!(\lambda+m)!(2l+1)}{(l-m)!(l+m)!(2\lambda+1)} \right\}^{1/2}$$

$$[(\lambda-l)!]^{-1}R^{\lambda-1}r_A^{-\lambda-1}Y_{\lambda m}(\hat{\mathbf{r}}_A) \qquad \text{for } r_A \geqq R$$

(2.11b)

It is customary to choose the coordinate axes for a diatomic molecule with the z-axis for each atom pointing inwards.

It follows that $\nabla_B^{\beta} = (-)^{\beta+1}\nabla_A^{\beta}$. In view of the finite sum (2.7) and the triangular condition which three combining angular momenta must satisfy, the infinite summations (2.11) reduce to finite sums in the final expression for the coulomb integral:

$$\langle \chi_A | \mathrm{h}'^{\mu}_{B1} | \chi_{A'} \rangle_C = \tfrac{1}{2}Z_A\delta(m_A, m_{A'}+\mu)R^{n_{A'}-1}$$

$$\times \sum_{\beta} \begin{pmatrix} 1 & 1 & 1 \\ \beta & -\beta-\mu & \mu \end{pmatrix} \sum_{\bar{n}\bar{l}} R^{\bar{n}} \sum_{\lambda} (-1)^{\beta+1} f^{\beta}_{\lambda 1}(\bar{n}\bar{l})$$

$$\times E_{n_{A'}+\bar{n}+\lambda}[(\zeta_A+\zeta_{A'})R] + f^{\beta}_{\lambda 2}(\bar{n}\bar{l})A_{n_{A'}+\bar{n}-\lambda-1}[(\zeta_A+\zeta_{A'})R]$$

(2.12)

where

$$f^{\beta}_{\lambda i}(\bar{n}\bar{l}) = (-1)^{m_A}k_{\bar{n}}k_{\bar{n}_A}[6(2\bar{l}+1)(2l_A+1)]^{\frac{1}{2}}\, b^{\beta}_{\lambda i}$$

$$\times c^{\beta+\mu}[n_{A'}l_{A'}m_{A'}|\bar{n}\bar{l}\ m_{A'}+\beta+\mu]$$

$$\times \begin{pmatrix} l_A & \lambda & \bar{l} \\ 0 & 0 & 0 \end{pmatrix} \begin{pmatrix} l_A & \lambda & \bar{l} \\ m_A & \beta & -m_{A'}-\beta-\mu \end{pmatrix} \qquad (2.13a)$$

$$b^0_{\lambda i} = \begin{cases} \lambda+1 & \text{for } i=1 \\ \lambda & \text{for } i=2 \end{cases} \qquad b^{\pm 1}_{\lambda i} = \{\tfrac{1}{2}\lambda(\lambda+1)\}^{\frac{1}{2}}. \qquad (2.13b)$$

The functions E and A are defined by Matcha and co-workers as:

$$E_j(t)=\int_0^1 x^j \exp(-tx)\mathrm{d}x \tag{2.14a}$$

and

$$A_j(t)=\int_1^\infty x^j \exp(-tx)\mathrm{d}x. \tag{2.14b}$$

The sum over λ in (2.12) progresses in steps of 2. The lower limit on λ in the sum involving the E_j function is $\lambda_{\min} = \max(|\beta|, |l_A - \bar{l}|)$ while for the sum involving the A_j functions it is $\lambda_{\min} = \max(1, |l_A - \bar{l}|)$.

Finally, for the hybrid integral, only ϕ integration can be performed analytically. The remaining integration is carried out in prolate spheroidal coordinates ξ and η by means of two-dimensional quadrature. For completeness, the final result turns out to be:

$$\begin{aligned}\langle\chi_A|\mathrm{h}'^{\mu}_{A1}|\chi_B\rangle_{\mathrm{H}} &= \tfrac{1}{2}Z_A\delta(m_A, m_B+\mu)\\ &\times\sum_r\sum_s \omega_r\omega_s \frac{\bar{\beta}(\bar{\beta}+1)}{(1-\bar{\beta}t_r)^2} U^{(2)}_{AB}\left(\frac{1+\bar{\beta}}{1-\bar{\beta}t_r}, \eta_s\right)\end{aligned} \tag{2.15}$$

where the two-centre function

$$\begin{aligned}U^{(2)}_{AB}(\xi,\eta) &= 2k_{n_A}(R/2)^{n_A-1}(\xi+\eta)^{n_A-2}P^{m_A}_{l_A}\left(\frac{1+\xi\eta}{\xi+\eta}\right)\\ &\times\exp(-\rho\xi-\tau\rho\eta)\sum_\beta (-1)^{\beta+1}\begin{pmatrix}1 & 1 & 1\\ \beta & -\beta-\mu & \mu\end{pmatrix}\\ &\times P_1^{-\beta}\left(\frac{1+\xi\eta}{\xi+\eta}\right)\sum_{\bar{n}\bar{l}} k_{\bar{n}}(R/2)^{\bar{n}}(\xi-\eta)^{\bar{n}}\\ &\times c^{\beta+\mu}[n_B l_B m_B|\bar{n}\bar{l}\ m_B+\beta+\mu]\ P^{m_B+\beta+\mu}_{\bar{l}}\left(\frac{1-\xi\eta}{\xi-\eta}\right)\end{aligned}$$

$$\text{with}\quad \rho=R(\zeta_A+\zeta_B)/2\ ,\ \tau\rho=R(\zeta_A-\zeta_B)/2 \tag{2.16a}$$

and the associated Legendre functions

$$P_l^m(x) = \frac{(-1)^m}{2^l l!}\left[\frac{2l+1}{2}\,\frac{(l-m)!}{(l+m)!}\right]^{1/2}(1-x^2)^{m/2} \times\left(\frac{\mathrm{d}}{\mathrm{d}x}\right)^{l+m}(x^2-1)^l \tag{2.16b}$$

The weights ω_r and ω_s of the Gauss-Legendre quadrature correspond to the grid-points labelled r and s which are the roots t_r and η_s of the Legendre polynomials in t and η of degree N and N' respectively. The somewhat arbitrary quantity $\bar{\beta}$ is chosen to transform the semi-infinite range $(1, \infty)$ of integration for ξ into a finite range $(-1, 1)$ for t, related to ξ by

$$t = (\xi - \bar{\beta} - 1)/(\bar{\beta}\xi)$$

In fact, $\bar{\beta} = (\xi_{\max} - 1)/(\xi_{\max} + 1)$ is so chosen that $U_{AB}^{(2)}$ is vanishingly small outside the spheroidal volume defined by $\xi_{\max}$.

Next, we come to the more difficult problem of the two-electron matrix elements of H'_{12}. As before, α^2 is understood and the prime denotes the space part only. This time there are five types of integral:

$\langle\chi_A(1)\chi_{A''}(2)\lvert H'^{\mu}_{12}\rvert\chi_{A'}(1)\chi_{A'''}(2)\rangle_0$	one-centre integral
$\langle\chi_A(1)\chi_{B'}(2)\lvert H'^{\mu}_{12}\rvert\chi_{A'}(1)\chi_B(2)\rangle_C$	coulomb integral
$\langle\chi_A(1)\chi_{B'}(2)\lvert H'^{\mu}_{12}\rvert\chi_B(1)\chi_{A'}(2)\rangle_E$	exchange integral
$\langle\chi_A(1)\chi_B(2)\lvert H'^{\mu}_{12}\rvert\chi_{A''}(1)\chi_{A'}(2)\rangle_{H1}$	hybrid integral (I)
$\langle\chi_B(1)\chi_A(2)\lvert H'^{\mu}_{12}\rvert\chi_{A''}(1)\chi_{A'}(2)\rangle_{H2}$	hybrid integral (II)

The last two integrals are distinct because H'^{μ}_{12} has no definite symmetry under the interchange of the two electrons. The operator

$$\begin{aligned} H'^{\mu}_{12} &= \{\nabla_1(1/r_{12})\times\mathbf{p}_1\}^{\mu} \\ &= \sqrt{6}\sum_{\beta\beta'}(-1)^{\mu}\nabla_i^{-\beta}(1/r_{12})\begin{pmatrix}1 & 1 & 1\\ \beta & -\beta' & \mu\end{pmatrix}\nabla_1^{\beta'}. \end{aligned}$$

The one-centre integral

$$\langle \chi_A(1)\chi_{A''}(2)|\mathrm{H}'^{\mu}_{12}|\chi_{A'}(1)\chi_{A'''}(2)\rangle_0 = \int \chi^*_{A''}(2)\chi_{A'''}(2)\mathrm{U}_A(2,\mu)\mathrm{d}V_2 \tag{2.17}$$

with

$$\mathrm{U}_A(2,\mu) = \int \chi^*_A(1)\mathrm{H}'^{\mu}_{12}\chi_{A'}(1)\mathrm{d}V_1$$

can be evaluated analytically. The intermediate steps are relatively straightforward though tedious and the final expression for (2.17) is given in Appendix 2.A.

The coulomb integral cannot be evaluated analytically.

$$\langle \chi_A(1)\chi_{B'}(2)|\mathrm{H}'^{\mu}_{12}|\chi_{A'}(1)\chi_B(2)\rangle_c = \int \Omega^*_B(2)\mathrm{U}_A(2,\mu)\mathrm{d}V_2 \tag{2.18}$$

where

$$\Omega_B(2) = \chi^*_B(2)\chi_{B'}(2)$$

Since Ω_B and $\mathrm{U}_A(2,\mu)$ are centred on two different nuclei it becomes convenient to express them in prolate spheroidal coordinates. After carrying out the ϕ integration analytically, the remaining integration is performed numerically. The final result is given in Appendix 2.B.

The exchange integral has been reduced to a one-dimensional numerical integration in the analysis of Matcha, Kern, and Schrader and an expression is given in Appendix 2.C to evaluate it. Matcha, Malli, and Milleur proposed a way of evaluating the hybrid integrals also by means of a one-dimensional quadrature. In this case, the spherical harmonics centred on B are expanded in terms of spherical harmonics centred on A and bipolar coordinates (r_A, r_B, ϕ) are found to be appropriate. An alternative form of the hybrid integral of type II is also encountered in practice,

$$\langle \chi_{A''}(1)\chi_A(2)|\mathrm{H}'^{\mu}_{12}|\chi_B(1)\chi_{A'}(2)\rangle.$$

This is readily dealt with using the already established machinery by noting that

$$\begin{aligned}&\langle m_{A''}(1)m_A(2)|\mathrm{H}'^{\mu}_{12}|m_B(1)m_{A'}(2)\rangle \\ &= (-1)^{m_{A''}-m_A+1}\langle -m_B(1)m_A(2)|\mathrm{H}'^{\mu}_{12}|-m_{A''}(1)m_{A'}(2)\rangle\end{aligned}$$

where $|-m_A\rangle = \chi_{n_A l_A - m_A}$ etc.

A series of programs has been developed at Oxford by Walker, Hall, Hammersley, Wilson, Cooper, and others to calculate the spin-orbit integrals using above methods. It is possible to use up to 100 bases, the highest being 4f, allowed molecular orbitals being of σ and π symmetry. For reasons to be explained later, the calculations are frequently limited to one-centre integrals.

2.5. The organization of calculations

Having calculated the atomic integrals it becomes necessary to carry out the transformation to molecular integrals:

$$\langle\phi_i(1)|\mathrm{h}'^{\mu}_1|\phi_j(1)\rangle = \sum_p\sum_q c^{\bullet}_{ip}c_{jq}\langle\chi_p(1)|\mathrm{h}'^{\mu}_1|\chi_q(1)\rangle \tag{2.19a}$$

$$\langle\phi_i(1)\phi_j(2)|\mathrm{H}'^{\mu}_{12}|\phi_k(1)\phi_l(2)\rangle = \sum_p\sum_q\sum_r\sum_s c^{\bullet}_{ip}c^{\bullet}_{jq}c_{kr}c_{ls} \times\langle\chi_p(1)\chi_q(2)|\mathrm{H}'^{\mu}_{12}|\chi_r(1)\chi_s(2)\rangle. \tag{2.19b}$$

The two-electron transformation deserves special consideration. Since it involves a large number of operations it is advisable to devise a strategy which accomplishes this task most efficiently. We abbreviate the matrix elements $\langle\chi_p(1)\chi_q(2)|\mathrm{H}'^{\mu}_{12}|\chi_r(1)\chi_s(2)\rangle$ as $\langle pq\|rs\rangle$ where the range of the basis functions $p, \ldots, s$ is n and the range of molecular orbitals i, j, k, l is m_i, m_j, m_k, m_l. A full transformation requires $m_i = m_j = m_k = m_l = n$. Usually, a full transformation is not required and Yoshimine has shown that if $m_i < m_j < m_k < m_l$ the most efficient course to follow is:

$$\begin{aligned}
\langle i\ q\|r\ s\rangle &= \sum_p c^{\bullet}_{ip}\ \langle p\ q\|r\ s\rangle \\
\langle i\ j\|r\ s\rangle &= \sum_q c^{\bullet}_{jq}\ \langle i\ q\|r\ s\rangle \\
\langle i\ j\|k\ s\rangle &= \sum_r c_{kr}\ \langle i\ j\|r\ s\rangle \\
\langle i\ j\|k\ l\rangle &= \sum_s c_{ls}\ \langle i\ j\|k\ s\rangle.
\end{aligned} \tag{2.20}$$

The total number of arithmetic operations is then $n^4m_i + n^3m_im_j + n^2m_im_jm_k + nm_im_jm_km_l$. If it is not possible to hold all the arrays in the fast memory – as is usually the case – the lists have to be broken up and direct-access storage devices (like disks) must be used. The array $\langle pq\|rs\rangle$ is held on a direct-access device in such a way that the nth set of disk record contains $\langle pq\|(rs)_n\rangle$ for all values of p and q, $(rs)_n$ being a particular combination of r and s. The first two of the transformations (2.20) are performed on each disk record so that the latter contains semi-transformed elements $\langle ij\|(rs)_n\rangle$. The semi-transformed records are sorted and stored according to $\langle (ij)_n\|rs\rangle$ for a particular combination $(ij)_n$ of i and j values corresponding to all values of r and s. Then the last two of the transformations (2.20) are carried out. Care must be exercised in view of the fact that $\langle pq\|rs\rangle \neq \langle qp\|sr\rangle$.

Use of various symmetry relations minimizes the number of distinct integrals which need be evaluated for

$$\langle m_i \; m_j|\mathrm{H}'^{\mu}_{12}|m_k \; m_l\rangle = (-1)^{m_j+m_l}\langle m_i \; -m_l|\mathrm{H}'^{\mu}_{12}|m_k \; -m_j\rangle \quad (2.21a)$$

$$= (-1)^{m_k+m_i+1}\langle -m_k \; m_j|\mathrm{H}'^{\mu}_{12}|-m_i \; m_l\rangle \quad (2.21b)$$

$$= -\langle -m_i \; -m_j|\mathrm{H}'^{-\mu}_{12}|-m_k \; -m_l\rangle \quad (2.21c)$$

where $|\pm m\rangle = |\lambda(n, l, \pm m)\rangle$. These relations are readily proved as follows:

$$\mathrm{H}'^{\mu}_{12} = \surd 6\sum_{\beta}(-1)^{\mu}\nabla_1^{-\beta}(1/r_{12})\nabla_1^{\beta+\mu}\begin{pmatrix} 1 & 1 & 1 \\ \beta & -\beta-\mu & \mu \end{pmatrix}$$

and

$$\langle m_i \; m_j|\mathrm{H}'^{\mu}_{12}|m_k \; m_l\rangle = \surd 6\sum_{\beta}(-1)^{\mu}\begin{pmatrix} 1 & 1 & 1 \\ \beta & -\beta-\mu & \mu \end{pmatrix}$$

$$\int\int \chi_i^*(1)\chi_j^*(2)\nabla_1^{-\beta}(1/r_{12})\nabla_1^{\beta+\mu}\chi_k(1)\chi_l(2)\mathrm{d}V_1\mathrm{d}V_2$$

Evaluating the integral by parts,

$$\int\int \chi_i^*(1)\chi_j^*(2)\nabla_1^{-\beta}(1/r_{12})\nabla_1^{\beta+\mu}\chi_k(1)\chi_l(2)\mathrm{d}V_1\mathrm{d}V_2$$

$$= -\int\int[\nabla_1^{\beta+\mu}\chi_i^*(1)]\chi_j^*(2)\nabla_1^{-\beta}(1/r_{12})\chi_k(1)\chi_l(2)\mathrm{d}V_1\mathrm{d}V_2$$

$$= -\int\int[\chi_k^*(1)]^*\chi_j^*(2)\nabla_1^{-\beta}(1/r_{12})\nabla_1^{\beta+\mu}[\chi_i^*(1)]\chi_l(2)\mathrm{d}V_1\mathrm{d}V_2$$

whence (2.21b) follows. Relation (2.21a) trivially follows from the fact that $Y_{l-m} = (-)^m (Y_{lm})^*$. Relation (2.2.1c) is arrived at using (2.21b) and the property $H'^{\mu}_{12} = (-)^{\mu+1} H'^{-\mu}_{12}$. A list of unique integrals for μ; $m = 0, \pm 1$ is given in Table 2.1.

Table 2.1

List of unique symmetry integrals for the operator H'^{μ}_{12}

$\mu = 0$	$\mu = +1$
$\langle \pi^+\sigma \mid \sigma\pi^+ \rangle$	$\langle \pi^+\sigma \mid \sigma\sigma \rangle$
$\langle \pi^+\sigma \mid \pi^+\sigma \rangle$	$\langle \sigma\pi^+ \mid \sigma\sigma \rangle$
$\langle \pi^+\pi^+ \mid \pi^+\pi^+ \rangle$	$\langle \pi^+\pi^+ \mid \pi^+\sigma \rangle$
$\langle \pi^+\pi^- \mid \pi^-\pi^+ \rangle$	$\langle \pi^+\pi^- \mid \sigma\pi^- \rangle$
	$\langle \pi^-\pi^+ \mid \pi^-\sigma \rangle$
	$\langle \pi^+\pi^- \mid \pi^-\sigma \rangle$
	$\langle \pi^-\pi^+ \mid \sigma\pi^- \rangle$

σ denotes $m = 0$. $\pi^{\pm}$ denotes $m = \pm 1$.

$\langle \pi^+\sigma \mid \sigma\pi^+ \rangle$ stands for $\langle \pi^+(1)\sigma(2) \mid H'^{\mu}_{12} \mid \sigma(1)\pi^+(2) \rangle$ and the notation is the same for the other matrix elements.

Slater's rules for the matrix elements of one- and two-body operators between (normalized) determinantal wavefunctions are summarized below. Between identical determinants ψ for a one-body operator Ω_1 they give

$$\langle \Psi | \Omega_1 | \Psi \rangle = \sum_i \langle i | \Omega_1 | i \rangle. \tag{2.22a}$$

For a two-body operator Ω_{12}

$$\langle \Psi | \Omega_{12} | \Psi \rangle = \sum_{i>j} \{ \langle i\ j | \Omega_{12} | i\ j \rangle - \langle i\ j | \Omega_{12} | j\ i \rangle \}. \tag{2.22b}$$

If the determinants differ by a single orbital, i.e. the single particle orbital i in ψ_i is replaced by a single-particle orbital j in ψ_j we have:

$$\langle \Psi_i | \Omega_1 | \Psi_j \rangle = \langle i | \Omega_1 | j \rangle \tag{2.23a}$$

and

$$\langle \Psi_i | \Omega_{12} | \Psi_j \rangle = \sum_l \{ \langle i\ l | \Omega_{12} | j\ l \rangle - \langle i\ l | \Omega_{12} | l\ j \rangle \}. \tag{2.23b}$$

If the determinants differ by two orbitals, ij in ψ_{ij}, and kl in ψ_{kl}

$$\langle \Psi_{ij} | \Omega_1 | \Psi_{kl} \rangle = 0 \tag{2.24a}$$

$$\langle \Psi_{ij}|\Omega_{12}|\Psi_{kl}\rangle = \langle i\ j|\Omega_{12}|k\ l\rangle - \langle i\ j|\Omega_{12}|l\ k\rangle. \quad (2.24b)$$

Matrix elements of Ω_1 and Ω_{12} between all other determinants vanish.

2.6. Discussion of some results

An extensive study of the spin–orbit coupling in Π states of several diatomic hydrides and oxides has been made. It has generally proved possible to get rather good agreement (~10 per cent) with experiment for the first-row hydrides. Oxides and fluorides turn out to be a little more difficult.

A substantial amount of empirical evidence has accumulated over the years to suggest that the contributions of the one- and the two-electron two-centre integrals tend to cancel each other. The individual contributions are small and the net contribution is frequently negligible (see Table 2.2). The interest in one-centre calculations stems from the considerable saving in computer time. The effect of including two-centre integrals for a number of diatomic species is shown in Table 2.3. The difference between the two sets of results is ~1 per cent for most first- and several second-row hydrides, and somewhat greater (~8 per cent) for oxides although BO and CO^+

Table 2.2

Contribution of one-centre and two-centre terms to calculated spin–orbit coupling constants

Molecule	(1)	(2)	(3)	(4)	*A*(calc.)	*A*(expt.)
BeH $A^2\Pi_r$	5.94	0.31	−3.91	−0.15	2.18	2.14
NH^+ $X^2\Pi_r$	135.57	1.43	−57.67	−1.14	78.19	77.8
OH^+ $A^3\Pi_i$	−124.57	−0.62	38.38	3.09	−83.72	−83.83
HF^+ $X^2\Pi_i$	−427.59	−1.40	134.56	1.08	−293.35	−292.86
CN $A^2\Pi_i$	−80.92	−10.24	34.02	8.27	−48.87	−52.20
NO $X^2\Pi_r$	200.80	−11.70	−79.20	3.88	113.78	124.2
CF $X^2\Pi_r$	114.02	−4.63	−46.86	1.27	63.80	77.1
PO $X^2\Pi_r$	281.78	−1.77	−71.05	1.03	209.99	224.0

All quantities are in cm^{-1} and calculated at, or near, R_e.

(1) One-electron one-centre contribution.
(2) One-electron two-centre contribution.
(3) Two-electron one-centre contribution.
(4) Two-electron two-centre contribution.

Table 2.3

Calculated spin-orbit coupling constants in cm^{-1}
Two-centre integrals are omitted from A_1 *but included in* A_2
R *denotes the interatomic separation in atomic units*
A *is experimental value*

Molecule	R	A_1	A_2	A
$BH^+(A^2\Pi_r{:}1{-}2\sigma^2, 1\pi)$	2.3743	14.0	13.8	14.0
$CH(X^2\Pi_r{:}1{-}3\sigma^2, 1\pi)$	2.124	28.8	28.5	28.0
$NH^-(X^2\Pi_i{:}1{-}3\sigma^2, 1\pi^3)$	1.9401	−55.6	−55.4	–
$FH^+(X^2\Pi_i{:}1{-}3\sigma^2, 1\pi^3)$	1.9843	−293.0	−293.2	–
$OH(X^2\Pi_i{:}1{-}3\sigma^2, 1\pi^3)$	1.8342	−140.6	−140.6	−139.7
$MgH(A^2\Pi_r{:}1{-}4\sigma^2, 1\pi^4, 2\pi)$	3.1739	53.6	53.4	35
$AlH^+(A^2\Pi_r{:}1{-}4\sigma^2, 1\pi^4, 2\pi)$	3.0066	91.4	91.1	108
$SiH(X^2\Pi_r{:}1{-}5\sigma^2, 1\pi^4, 2\pi)$	2.8724	127.1	126.8	142
$PH^-(X^2\Pi_i{:}1{-}5\sigma^2, 1\pi^4, 2\pi^3)$	2.668	−173.5	−173.1	–
$SH(X^2\Pi_i{:}1{-}5\sigma^2, 2\pi^4, 2\pi^3)$	2.5417	−352.4	−351.6	−382.4
$CO(A^3\Pi_r{:}1{-}4\sigma^2, 5\sigma, 1\pi^4, 2\pi)$	2.2853	45.2	41.6	41.5
$BO(A^2\Pi_i{:}1{-}5\sigma^2, 1\pi^3)$	2.5557	−119.5	−119.6	−116.7
$BO(C^2\Pi_r{:}1{-}4\sigma^2, 1\pi^4, 2\pi)$	2.4937	39.5	35.0	46.4
$CO^+(A^2\Pi_i{:}1{-}5\sigma^2, 1\pi^3)$	2.3501	−135.6	−134.4	−117.5
$CO^+(C^2\Pi_r{:}1{-}4\sigma^2, 1\pi^4, 2\pi)$	2.3501	143.7	138.1	–
$BeF(A^2\Pi_r{:}1{-}4\sigma^2, 1\pi^4, 2\pi)$	2.714	–	13.5	21.8

exhibit very small effects. CH shows a somewhat large (5 per cent) effect.

To refine the agreement with experiment one must take greater account of electron correlation. It is most convenient to work with SCF–CI wavefunctions. Unfortunately, addition of more configuration state functions (CSFs) does not always result in uniform improvement and the convergence is often very slow. The calculated values of A in the $^2\Delta$ states of CF and NH^+ are given in Table 2.4. The computer time increases dramatically with the number of CSFs, much of the increase in time being in the sorting of symbolic matrix elements. For instance, dealing with 470 CSFs in the molecule SiH involves the sorting of over a million matrix elements and takes about 13 minutes on an IBM 360/195 computer.

Unrestricted Hartree–Fock (UHF) wavefunctions are flexible enough to accommodate the fact that the radial wavefunctions of

Table 2.4

Effect of using configuration interaction on the calculated value of the coupling constant A Only one-centre integrals are included

Molecule	R(a.u.)	HF value (cm^{-1})	CI value (cm^{-1})	No. of CSFs
$NH^+(X^2\Pi)$	2.05	84.17	77.91	305
$NH^+(B^2\Delta)$	2.05	−4.41	−2.88	276
$CF(B^2\Delta)$	2.4	−2.39	−6.94	162
$CF(B^2\Delta)$	2.4	−2.39	−4.07	405
$SiH(X^2\Pi)$	2.87	126.8	121.97	16

two electrons differing only in spin are slightly different. This is known as the core-polarization effect. UHF wavefunctions are difficult to obtain for molecules using STO bases. Some calculations have used the known core-polarization effect in atoms to estimate its impact on molecules and have succeeded in reducing the discrepancy with experiment (see Tables 2.5 and 2.6).

Table 2.5

Effects of core polarization on contributions to atomic spin–orbit coupling constants (cm^{-1})†

	Na§	Al	Si	S	Cl
Change in 3p contribution	0	1.4	0	−28	−15
Resultant 2p contribution‡	3.2	16.6	25.6	48	64

†Lefèbvre-Brion, H., Wajsbaum, J., and Bessis, N., in *La structure hyperfine des atomes et des molecules*, Centre National de la Recherche Scientifique, Paris (1967).

‡Without core polarization this contribution is zero.

§Na, ^2P.

Spin–orbit coupling energy varies with internuclear separation for two distinct reasons. Firstly there is the variation of spin–orbit coupling matrix elements. In the case of the $X^2\Pi_g$ state of O_2^+ the change in the $1\pi_g$ orbital is responsible for the change in A. On the other hand, it is the different relative importance of certain configuration state functions in the configuration interaction wavefunction for the $A^2\Pi_u$ state of O_2^+ that causes the variation of A in this state.

Table 2.6
Spin–orbit coupling constants including 'core polarization'

Molecular state	Calc† (cm^{-1}) 1	2	Obs‡ (cm^{-1})
MgH $A^2\Pi_r$	26.8	36.4	35
AlH^+ $A^2\Pi_r$	92.9	110.9	108
SiH $X^2\Pi_r$	129.5	155.1	142
PH^- $X^2\Pi_i$	175.3	211.7	—
SH $X^2\Pi_i$	355.7	375.7	382
PO $X^2\Pi_r$	153.7	190.1	224
ClO $X^2\Pi_i$	209.1	258.1	289
MgF $A^2\Pi_r$	31.3	40.9	37

†Columns: 1, Without 'core polarization' correction; 2, with correction.
‡Herzberg, G., *Spectra of diatomic molecules*. D. Van Nostrand, New York (1950).

The second effect is the change in the dependence of the spin–orbit coupling energy on the spin–orbit coupling constant. From the Landé interval rule (see Chapter 1) the energy separation between the J and $(J-1)$ levels arising from the same atomic term is JA. The spin–orbit coupling energy in a diatomic molecule is $A\Lambda\Sigma$ so that the energy separation between the Ω and $(\Omega-1)$ levels arising from the same term is $2A\Lambda\Sigma$. Consider a $^2\Pi$ molecule dissociating to a closed shell $^1S_0$ atom and to an open shell ^2P atom. In the molecule, the splitting between the $^2\Pi_{1/2}$ and $^2\Pi_{3/2}$ levels is A – even at infinite bond length. Obviously, the system at infinite separation is not a molecule but is two separate atoms with spin–orbit splitting $\frac{3}{2}A$. Clearly, there is a transition region in which the spin–orbit coupling energy changes between these two extremes. Systems such as $HeNe^+$ are particularly interesting: it is unlikely that there would be a large interaction between the two atoms except at very short internuclear separation and so the spin–orbit coupling energy might be more representative of Ne^+ than of $HeNe^+$. Measurement of this quantity may soon be possible; there have already been suggestions that close to R_e, the system is in the transition region mentioned above.

Finally, the enigma of the first excited $A^2\Pi$ state in BeF and MgF must be mentioned. Experimental evidence remains inconclusive. Some experiments suggest a regular configuration. For example, the transition $C^2\Sigma^+ \rightarrow A^2\Pi$ observed in emission spectra coupled with the knowledge that the Rydberg state $C^2\Sigma^+$ contains the $\sigma^2\pi^4$ core implies that $A\,^2\Pi$ contains the same core. If not, the transition would necessitate a two-electron jump and would therefore be forbidden. On the other hand, analysis of the absorption spectra suggests an inverted configuration. Theoretical calculations of the spin-orbit coupling constant suggest a regular structure; the inverted configuration does not even appear to have a minimum in its potential curve. As against this, a study of the Λ-doubling effect strongly suggests an inverted character for the state. To date, the question remains unresolved.

Further reading

1. Slater, J. C., *Quantum theory of atomic structure*, Volumes I and II. McGraw-Hill, New York (1960).
2. Herzberg, G., *Molecular spectra and molecular structure I. Spectra of diatomic molecules*. Van Nostrand, New York (1950).
3. Matcha, R. L., Kern, C. W., and Schrader, D. M., Fine structure studies of diatomic molecules: two-electron spin-spin and spin-orbit integrals. *J. chem. Phys.* **51**, 2152 (1969).
4. Matcha, R. N., Malli, G., and Milleur, M. B., Two-centre two-electron spin-spin and spin-orbit hybrid integrals. *J. chem. Phys.* **56**, 5982 (1972).
5. Huber, K. P. and Herzberg, G., *Molecular spectra and molecular structure IV. Constants of diatomic molecules*. Van Nostrand Reinhold, New York (1979). This reference together with reference 2 above gives all the relevant experimental data.
6. Veseth, L., Spin-orbit and spin-other-orbit interaction in diatomic molecules. *Theoret. Chim. Acta (Berl.)* **18**, 368 (1970).
7. Raftery, J. and Richards, W. G., On the variation of the spin-orbit coupling in O_2^+. *J. chem. Phys.* **62**, 3184 (1975).

APPENDIX 2.A.
TWO-ELECTRON ONE-CENTRE INTEGRALS

$$U_A(2,\mu) = \sqrt{(2\pi)}\, r_{A2}^{n_A-1} \sum_{\beta \bar{l}_A} \sum_{\kappa} \sum_{\bar{n}_A} (-1)^{m_A+\beta+\mu+\kappa}\, r_{A2}^{\bar{n}_A}$$
$$Y_{kq'}(\hat{\mathbf{r}}_{A2})\big[J^{\mu}_{\kappa\beta}(\kappa-1,q',\bar{n}_A,\bar{l}_A)E_{n_A+\bar{n}_A+\kappa}(1)$$
$$+J^{\mu}_{\kappa\beta}(\kappa+1,q',\bar{n}_A,\bar{l}_A)A_{n_A+\bar{n}_A+\kappa}(2)\big],$$

where

$$q'=m_{A'}-m_A+\mu$$
$$\kappa_{\min}=\max\{|l_A-\bar{l}_A|+1,|p+q'|+1\},$$

and

$$\kappa_{\max}=l_A+\bar{l}_A+1$$

for the sum involving functions E_j while $\kappa_{\min} = \max(|l_A - \bar{l}_A| - 1, |\beta + q'| - 1)$ and $\kappa_{\max} = l_A + \bar{l}_A - 1$ for the sum involving functions A_j, the summation proceeding in steps of 2 in both cases. The auxiliary function

$$J^{\mu}_{\kappa\beta}(\lambda,q,\bar{n}_A,\bar{l}_A) = \bar{\kappa}_A\{12[\kappa+\delta(\lambda,\kappa+1)](2l_A+1)(2\bar{l}_A+1)(2\lambda+1)\}^{1/2}$$
$$c^{\beta+\mu}[n_{A'}l_{A'}m_{A'}|\bar{n}_A\bar{l}_A\, m_{A'}+\beta+\mu]\begin{pmatrix} 1 & 1 & 1 \\ \beta & -\beta-\mu & \mu \end{pmatrix}$$
$$\times\begin{pmatrix} \kappa & 1 & \lambda \\ q & \beta & -q-\beta \end{pmatrix}\begin{pmatrix} l_A & \lambda & \bar{l}_A \\ m_A & \beta+q & -m_{A'}-\beta-\mu \end{pmatrix}$$
$$\times\begin{pmatrix} l_A & \lambda & \bar{l}_A \\ 0 & 0 & 0 \end{pmatrix},$$

where

$$\bar{k}_A = k(n_A, \bar{n}_A; \zeta_A, \zeta_{A'})$$

and
$$k(i,j;x,y) = (2x)^{i+1/2}(2y)^{j+1/2}[(2i)!(2j)!]^{-1/2}.$$

Substituting this expression for $U_A(2, \mu)$, the final result turns out to be

$$\begin{aligned}\langle H'^{\mu}_{12}\rangle_0 &= \delta(m_{A''} - m_{A'''}, m_{A'} - m_A + \mu) \\ &\sum_{\beta \bar{l}_A} \sum_{\kappa} \sum_{\bar{n}_A} [G^{\mu}_{\kappa\beta}(\kappa - 1, q', \bar{n}_A, \bar{l}_A) H^{\kappa}_{\zeta_{A1}\zeta_{A2}}(\bar{n}_2, \bar{n}_1) \\ &+ G^{\mu}_{\kappa\beta}(\kappa + 1, q', \bar{n}_A, \bar{l}_A) H^{\kappa}_{\zeta_{A1}\zeta_{A2}}(\bar{n}_1, \bar{n}_2)]\end{aligned} \tag{2.A.1}$$

with

$$\bar{n}_1 = n_{A''} + n_{A'''} - 1, \qquad \bar{n}_2 = n_A + n_{A'},$$

$$H^{\kappa}_{ab}(N, \bar{N}) = a^{-N-k} b^{-\bar{N}+\kappa+1} \bar{H}(p, q, s),$$

$$p = N + \kappa,$$

$$q = \bar{N} - \kappa - 1,$$

$$s = a/b, \quad \text{and}$$

$$\bar{H}(p, q, s) = (p-1)!(q-1)! B_{1/(1+s)}(p, q),$$

$$B_{1/(1+s)}(p, q) = \frac{(p+q-1)!}{(p-1)!} \sum_i (-1)^i (1+s)^{-(p+i)} / [(p+i)(q-i-1)! i!]$$

being the incomplete beta function.

Also,

$$\begin{aligned}G^{\mu}_{\kappa\beta}(\lambda, q', \bar{n}_A, \bar{l}_A) &= (-1)^{m_A + m_{A''} + \beta + \kappa + \mu} k(n_{A''}, n_{A'''}; \zeta_{A''}, \zeta_{A'''}) \\ &J^{\mu}_{\kappa\beta}(\lambda, q', \bar{n}_A, \bar{l}_A)[\tfrac{1}{2}(2l_{A''} + 1)(2\kappa + 1)(2l_{A'''} + 1)]^{\frac{1}{2}} \\ &\begin{pmatrix} l_{A''} & \kappa & l_{A'''} \\ m_{A''} & -q' & -m_{A'''} \end{pmatrix} \begin{pmatrix} l_{A''} & \kappa & l_{A'''} \\ 0 & 0 & 0 \end{pmatrix}.\end{aligned}$$

The limits on the sums in (2.A.1) are

$$\kappa_{\min} = \max\{|l_A - \bar{l}_A| + 1, |\beta + q'| + 1, |m_{A''} - m_{A'''}|, |l_{A''} - l_{A'''}|\},$$

$$\kappa_{\max} = \min\{l_A + \bar{l}_A + 1, l_{A''} + l_{A'''}\}$$

for the first sum while

$$\kappa_{\min} = \max\{|l_A - \bar{l}_A| - 1, |\beta + q'| - 1, |m_{A''} - m_{A'''}|, |l_{A''} - l_{A'''}|\}$$

$$\kappa_{\max} = \min\{l_A + \bar{l}_A - 1, l_{A''} + l_{A'''}\}$$

for the second sum, κ progressing in steps of 2.

APPENDIX 2.B.
TWO-ELECTRON COULOMB INTEGRALS

In prolate spheroidal coordinates,

$$\mathrm{U}_A(2,\mu)=\mathrm{U}_A(\xi,\eta)\exp[i(m_{A'}-m_A+\mu)\phi]$$

with $\xi=(r_{A2}+r_{B2})/R,$ $\eta=(r_{A2}-r_{B2})/R,$ and

$$\mathrm{U}_A(\xi,\eta)=[\tfrac{1}{2}R(\xi+\eta)]^{n_A-1}\sum_{\beta\bar{l}_A}\sum_{\kappa}\sum_{\bar{n}_A}(-1)^{m_A+\beta+\mu+\kappa}$$

$$[\tfrac{1}{2}R(\xi+\mu)]^{\bar{n}_A}\mathrm{P}_{\kappa}^{q'}\left(\frac{1+\xi\eta}{\xi+\eta}\right)$$

$$\{\mathrm{J}^{\mu}_{\kappa\beta}E_{n_A+\bar{n}_A+\kappa-1}[\tfrac{1}{2}R(\zeta_A+\zeta_{A'})(\xi+\eta)]$$

$$+\mathrm{J}^{\mu}_{\lambda\beta}A_{n_A+\bar{n}_A-\kappa-2}[\tfrac{1}{2}R(\zeta_A+\zeta_{A'})(\xi+\eta)]\}.$$

Here
$$\mathrm{P}_l^m(x)=\left[\frac{(2l+1)(l-m)!}{2(l+m)!}\right]^{1/2}P_l^m(x)$$

is the orthonormalized associated Legendre function while

$$P_l^m(x)=\frac{(-1)^m}{2^l l!}(1-x^2)^{m/2}\left(\frac{\mathrm{d}}{\mathrm{d}x}\right)^{l+m}(x^2-1)^l.$$

All other functions and symbols are the same as in §2.4 and Appendix 2.A. Also the same summation steps and limits apply as before.

After carrying out the ϕ integration analytically, the two-electron Coulomb integral reduces to a two-dimensional integration

$$\langle \chi_A(1)\chi_{B'}(2)|\mathrm{H}'^{\mu}_{12}|\chi_{A'}(1)\chi_B(2)\rangle_c = \delta(m_{A'}-m_A+\mu, m_B-m_{B'})$$
$$\times\int_{-1}^{1} d\eta \int_{1}^{\infty} d\xi\, \mathrm{U}_A(\xi,\eta)\Omega_B(\xi,\eta)(\tfrac{1}{2}R)^3(\xi^2-\eta^2).$$

This can be performed using Gauss–Legendre quadrature as described in connection with one-electron hybrid integral in §2.4. The final result is

$$\langle \mathrm{H}'^{\mu}_{12}\rangle_c = \delta(m_{A'}-m_A+\mu, m_B-m_{B'})$$
$$\times\sum_r\sum_s \left[\omega_r\omega_s(\tfrac{1}{2}R)^3 \frac{\bar{\beta}(\bar{\beta}+1)}{(1-\bar{\beta}\,t_r)^2}(\xi^2-\eta_s^2)\mathrm{U}_A(\xi,\bar{\eta}_s)\right]$$
$$\times\,\Omega_B(\xi,\bar{\eta}_s), \qquad \xi = \frac{1+\bar{\beta}}{(1-\bar{\beta}\,t_r)}$$

r and s being the grid points, ω_r and ω_s their respective weights, and $\bar{N}$ and $\bar{N}'$ determining the size of the grid mesh.

APPENDIX 2.C
TWO-ELECTRON EXCHANGE INTEGRALS

The exchange integral

$$\langle \chi_A(1)\chi_{B'}(2)|\mathrm{H}'^{\mu}_{12}|\chi_B(1)\chi_{A'}(2)\rangle_{\mathrm{E}}$$

$$= -\sqrt{6}\sum_{\beta\beta'}(-1)^{\mu}\begin{pmatrix} 1 & 1 & 1 \\ \beta & -\beta' & \mu \end{pmatrix}$$

$$\times \int\int \Omega^{*}_{A'B'}(2)\chi^{*}_{A}(1)\nabla_2^{-\beta}(1/r_{12})\nabla_1^{\beta'}\chi_B(1)\,\mathrm{d}V_1\mathrm{d}V_2$$

becomes, after some manipulations,

$$= (-1)^{m_B+m_{B'}}\langle \mathrm{I}^{\mu}_{12}(-m_B,-m_{B'})\rangle$$

where $\langle \mathrm{I}^{\mu}_{12}\rangle = \mathrm{I}^{\mu}_{12}(AB|AB)+\mathrm{I}^{\mu}_{12}(AB|BA)$,

$$\mathrm{I}^{\mu}_{12}(AB|CD) = \sqrt{6}\sum_{\beta\beta'}(-1)^{\mu}\begin{pmatrix} 1 & 1 & 1 \\ \beta' & -\beta & -\mu \end{pmatrix}$$

$$\times \int\int r_{12}^{-1}[\chi^{*}_{A}(1)\nabla_1^{-\beta}\chi^{*}_{B}(1)][\chi_{C'}(2)\nabla_2^{\beta'}\chi_{D'}(2)]\,\mathrm{d}V_1\mathrm{d}V_2$$

and $\mathrm{I}^{\mu}_{12}(-m_B,\ -m_{B'})$ implies that $\mathrm{I}^{\mu}_{12}(AB\,|AB)$ and $\mathrm{I}^{\mu}_{12}(AB\,|BA)$ are evaluated with $-m_B$ and $-m_{B'}$ in place of m_B and $m_{B'}$, respectively.

Matcha, Kern, and Schrader show that this reduces to

$$\langle \mathsf{I}_{12}^{\mu} \rangle = (-1)^{\mu+1}[4\sqrt{6}/R]\mathsf{K}_{AB}\mathsf{K}_{A'B'}\delta(M, M'+\mu)$$

$$\times \sum_{\beta} \begin{pmatrix} 1 & 1 & 1 \\ \beta+\mu & -\beta & -\mu \end{pmatrix} \sum_{\lambda} \mathsf{I}_{\lambda}^{\beta,\beta+\mu}$$

β ranging from -1 to $+1$ and λ from M' to ∞. Here

$$\mathsf{K}_{AB} = (-1)^{M}(2\zeta_A)^{n_A+\frac{1}{2}}(2\zeta_B)^{n_B+\frac{1}{2}}[(2n_A)!(2n_B)!]^{-1/2}$$

$$\times (R/2)^{N+1} \left[\frac{(2l_A+1)(2l_B+1)(l_A-m_A)!(l_B-m_B)}{4(l_A+m_A)!(l_B+m_B)!} \right]^{1/2}$$

$$M = m_A + m_B \qquad N = n_A + n_B$$

$$M' = m_{A'} + m_{B'} \qquad N' = n_{A'} + n_{B'}$$

and the Neumann expansion of $1/r_{12}$ in spherical polar coordinates is used. The definition of $\mathsf{I}^{\beta,\beta+\mu}$ and its evaluation are rather involved and are best described in the original paper of Matcha, Kern, and Schrader (1969).

3

OFF-DIAGONAL SPIN–ORBIT EFFECTS IN DIATOMIC MOLECULES

3.1. Introduction

We have already seen how diagonal matrix elements of H_{SO} can be calculated and we have examined their importance in a variety of situations. We shall now consider the off-diagonal matrix elements of this operator and we shall find that these have a large number of important uses.

The operator H_{SO} is, of course, the same as described in Chapter 2 but the selection rules are somewhat different. These may be summarized as follows:

$$\Delta S = 0, \pm 1 \quad \Delta J = 0$$

$$\Delta \Lambda = 0, \pm 1 \quad \Delta \Sigma = -\Delta \Lambda \quad \Delta \Omega = 0$$

$$g \longleftrightarrow g \quad u \longleftrightarrow u \quad g \not\longleftrightarrow u$$

$$+ \longleftrightarrow - \quad + \not\longleftrightarrow + \quad - \not\longleftrightarrow -$$

The main effects of off-diagonal spin–orbit coupling are seen in Π and Σ states. A much studied phenomenon is the Λ-type doubling observed in $^2\Pi$ states: this is the lifting of the degeneracy of pairs of levels with different parity. It arises because of interactions with $^2\Sigma^+$ and $^2\Sigma^-$ through matrix elements of the spin–orbit coupling and Coriolis parts of the full Hamiltonian.

Similarly, spin–rotation splitting in $^2\Sigma$ states occurs because H_{SO} indirectly couples the spin angular momentum to the rotational angular momentum of the molecule. We shall also discover that matrix elements of H_{SO} appear not only in the explanations of various perturbations observed in electronic spectra but also in discussions of celestial masers.

3.2. Λ-doubling

3.2.1. *Introduction*

Λ-doubling arises from the mixing of Π and Σ states and may be regarded as a breakdown of the Born–Oppenheimer approximation. The motions of the electrons and nuclei may not be treated as totally independent.

Λ-doubling may be very crudely pictured in terms of 'electron slip' but it must be stressed that this picture must not be taken too seriously. In a rotating molecule, the electrons may lag behind the motion of the nuclear framework: the electrons and the nuclear framework thus have slightly different angular momenta. The diagram below (Fig. 3.1) shows the effect of rotation on the otherwise degenerate π^+ and π^- orbitals. Whereas the π^- orbital is scarcely affected, the nuclei in the case of the π^+ orbital can come into a region of non-zero electron density (as in σ orbitals). There is a lifting of the degeneracy of orbitals with the same value of $|\Omega|$, a blurring of the distinction between σ and π molecular orbitals, and a removal of the orthogonality of Σ and Π molecular states. It is to be expected from this picture that the Λ-doublet splitting would increase with rate of rotation (and thus with J).

Spectroscopists usually fit the Λ-doublet splitting observed in different rotational levels to an expressing containing J and two parameters $(\frac{1}{2}p + q)$ and q. Even very crude calculations of these parameters are of value since they may give some indication of the origins of the magnitudes of measured properties. In particular, the relative signs of p and q may be used to determine whether a state is regular or inverted. The first excited state of BeF is $A\,^2\Pi$ with a negative q and a positive p; this suggests that the state is inverted. This contradicts other evidence and so calculations would be valuable.

More accurate calculations allow estimates of molecular constants in cases where the experimental method is difficult. Electronic spectroscopy has particular difficulties in the case of very reactive species whose lifetimes may be too short for very high-resolution studies. The importance of even more accurate *ab initio* calculations arises from the radio-frequency, electric-dipole allowed transitions between the components of the Λ-doublet.

The interstellar medium often hinders the study of distant objects but is itself an important source of information concerning the evolution of the universe. Most of what is known about the inter-

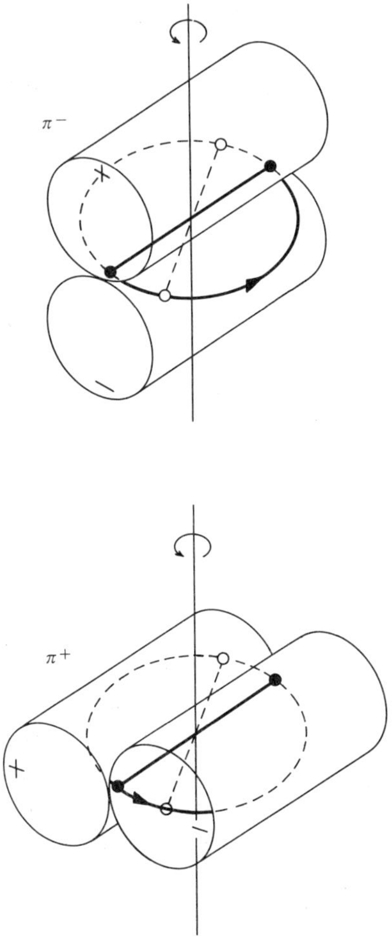

Fig. 3.1. A schematic representation of the mechanism for Λ-doubling in a diatomic molecule. On rotation if the electrons lag slightly, in π^- the nuclei will remain in a nodal plane of the orbital while in π^+ the nuclei will come into a region of non-zero electron density.

stellar dust clouds, which are believed to be the precursors of new stars, is based on recent observations using radiospectroscopy. A large variety of molecules has been observed in the dust clouds where they are sufficiently protected from photolysing radiation to have long lifetimes and quite large steady-state concentrations. Many interstellar molecules have been detected through the electric-dipole allowed spontaneous emission of radiation due to rotational transitions. However, for the lightest species, the frequency of even the lowest rotational transitions are outside the range detectable by radio telescopes. Molecules such as OH have been observed because of transitions between components of Λ-doublets. Similarly, after several years of searching, CH was observed at a frequency closer to that predicted by *ab initio* calculation than to terrestrial spectroscopic estimates.

Electronic spectroscopy is generally not carried out under the conditions that prevail in the molecular clouds: very low temperatures and densities. Even if high-resolution spectra can be produced, a large amount of work must go into extrapolating quantities relating to excited rotational levels. This fitting procedure may involve large uncertainties. Furthermore, different experimentalists use very different extrapolation routines. It is interesting to note that the ion N_2H^+ is so reactive that its study is far easier in space than in a laboratory.

Since rather an accurate frequency prediction is required before a search can be made for a new molecule such as SiH or NH^+, it is clear that Λ-doubling calculations are of considerable importance. Study of interstellar molecules not only gives information concerning temperatures, concentrations, motions, etc. of individual clouds but also estimates of relative abundances of atoms. In particular the H:D abundance ratio gives some idea of the date of the 'big bang'. Clearly it is important to try and understand the processes by which these molecules are formed. Indeed, one school of thought believes in the 'panspermia' theory which proposes that life originated in space.

3.2.2. *The mixing of states*

Having seen a qualitative picture of the origin of Λ-doubling and having considered some of the reasons why it is important, we shall proceed to a formal treatment of this phenomenon. Off-diagonal matrix elements of H_{SO} and BJ^+L^- cause the mixing of states with different $|\Lambda|$ values.

First consider the perturbation of a single vibronic level of a $^2\Pi$ state by a single vibronic level of a $^2\Sigma^+$ state. The full molecular Hamiltonian may be written:

$$H = H^{(0)} + H^{(1)}$$

where $H^{(0)}$ is the vibronic Hamiltonian and $H^{(1)}$ is the perturbation $H_{SO} + H_{rot}$. Much of the treatment of Λ-doubling assumes Hund's case (a), but arguments based on case (b) must produce equivalent results: we shall abbreviate a Hund's case (a) wavefunction as $|\Lambda S\Sigma J\rangle$.

The rotational part of the perturbation may be expanded:

$$\begin{aligned} \mathrm{H_{rot}} &= B[\mathbf{J}-\mathbf{L}-\mathbf{S}]^2 \\ &= B[\mathrm{J}_x-\mathrm{L}_x-\mathrm{S}_x]^2 + B[\mathrm{J}_y-\mathrm{L}_y-\mathrm{S}_y]^2 \\ &= B(\mathrm{J}^2-\mathrm{J}_z^2) + B(\mathrm{S}^2-\mathrm{S}_z^2) + B(\mathrm{L}^2-\mathrm{L}_z^2) \\ &\quad + B(\mathrm{L}^+\mathrm{S}^-+\mathrm{L}^-\mathrm{S}^+) - B(\mathrm{J}^+\mathrm{L}^-+\mathrm{J}^-\mathrm{L}^+) \\ &\quad - B(\mathrm{J}^+\mathrm{S}^-+\mathrm{J}^-\mathrm{S}^+). \end{aligned}$$

To find the perturbed energies, we must construct the perturbation matrix for the space spanned by:

$$\langle{}^2\Pi_{3/2}|,\ \langle{}^2\Pi_{1/2}|,\ \langle{}^2\Sigma^+_{1/2}|,\ \langle{}^2\Pi_{-3/2}|,\ \langle{}^2\Pi_{-1/2}|,\ \text{and}\ \langle{}^2\Sigma^+_{-1/2}|.$$

Many of the matrix elements are zero because of selection rules – some of the various non-zero terms are listed below. Note that the term $(\mathrm{L}^2 - \mathrm{L}_z^2)$ that arises is neglected on the grounds that it is a very small, diagonal term independent of J or S.

$$\begin{aligned}
\langle{}^2\Pi_{3/2}|\mathrm{H}^{(1)}|{}^2\Pi_{3/2}\rangle &= B_\Pi[J(J+1)-\tfrac{7}{4}] + \tfrac{1}{2}A_\Pi \\
&= \alpha \\
\langle{}^2\Pi_{3/2}|\mathrm{H}^{(1)}|{}^2\Pi_{1/2}\rangle &= -B_\Pi[(J+\tfrac{3}{2})(J-\tfrac{1}{2})]^{\frac{1}{2}} \\
&= \beta \\
\langle{}^2\Pi_{1/2}|\mathrm{H}^{(1)}|{}^2\Pi_{1/2}\rangle &= B_\Pi[J(J+1)+\tfrac{1}{4}] - \tfrac{1}{2}A_\Pi \\
&= \delta \\
\langle{}^2\Pi_{3/2}|\mathrm{H}^{(1)}|{}^2\Sigma^+_{1/2}\rangle &= -[(J+\tfrac{3}{2})(J-\tfrac{1}{2})]^{\frac{1}{2}}\ \langle{}^2\Pi_{3/2}|B(\mathrm{J}^+\mathrm{L}^-+\mathrm{J}^-\mathrm{L}^+)|{}^2\Sigma^+_{1/2}\rangle \\
&= \gamma \\
\langle{}^2\Pi_{1/2}|\mathrm{H}^{(1)}|{}^2\Sigma^+_{1/2}\rangle &= \langle{}^2\Pi_{1/2}|B(\mathrm{L}^+\mathrm{S}^-+\mathrm{L}^-\mathrm{S}^+)|{}^2\Sigma^+_{1/2}\rangle + \langle{}^2\Pi_{1/2}|\mathrm{H_{SO}}|{}^2\Sigma^+_{1/2}\rangle \\
&= \mu \\
\langle{}^2\Pi_{1/2}|\mathrm{H}^{(1)}|{}^2\Sigma^+_{-1/2}\rangle &= -(J+\tfrac{1}{2})\langle{}^2\Pi_{1/2}|B(\mathrm{J}^+\mathrm{L}^-+\mathrm{J}^-\mathrm{L}^+)|{}^2\Sigma^+_{-1/2}\rangle \\
&= -\nu \\
\langle{}^2\Sigma^+_{1/2}|\mathrm{H}^{(1)}|{}^2\Sigma^+_{-1/2}\rangle &= -B_\Sigma(J+\tfrac{1}{2}) \\
&= \theta \\
\langle{}^2\Sigma^+_{1/2}|\mathrm{H}^{(1)}|{}^2\Sigma^+_{1/2}\rangle &= B_\Sigma[J(J+1)+\tfrac{1}{4}] - \Delta E \\
&= \phi
\end{aligned}$$

We may thus quickly proceed to the perturbation matrix:

$$\mathbf{P} = \begin{pmatrix} \alpha & \beta & \gamma & 0 & 0 & 0 \\ \beta & \delta & \mu & 0 & 0 & -\nu \\ \gamma & \mu & \phi & 0 & -\nu & \theta \\ 0 & 0 & 0 & \alpha & \beta & \gamma \\ 0 & 0 & -\nu & \beta & \delta & -\nu \\ 0 & -\nu & \theta & \gamma & -\nu & \phi \end{pmatrix}$$

The problem may be simplified by block-diagonalizing **P**. This is achieved by taking linear combinations:

$$|\Lambda S\Sigma J\pm\rangle = 1/\sqrt{2}(|\Lambda S\Sigma J\rangle \pm |-\Lambda S-\Sigma J\rangle).$$

The result of this is two 3 × 3 blocks (corresponding to the + and − combinations) for the spaces spanned by $|^2_\pm\Pi_{3/2}\rangle$, $|^2_\pm\Pi_{1/2}\rangle$, and $|^2_\pm\Sigma^+_{1/2}\rangle$.

$$\mathbf{P}_\pm = \begin{pmatrix} \alpha & \beta & \gamma \\ \beta & \delta & \mu\mp\nu \\ \gamma & \mu\mp\nu & \phi\pm\theta \end{pmatrix}.$$

The multiplet splitting between $^2\Pi_{3/2}$ and $^2\Pi_{1/2}$, the Λ-splitting, and the rotational constant B_Σ, are all small compared with the energy separation between the $^2\Pi$ and $^2\Sigma$ potential curves. We may thus approximate the determinants to

$$\begin{vmatrix} \alpha-\mathrm{T}_\pm & \beta & \gamma \\ \beta & \delta-\mathrm{T}_\pm & \mu\mp\nu \\ \gamma & \mu\mp\nu & -\Delta E \end{vmatrix} = 0.$$

Expanding the determinant and collecting terms

$$-\Delta E T_\pm^2 + T_\pm[(\alpha+\delta)\Delta E + [(\mu\mp\nu)^2+\gamma^2]$$
$$+[-\alpha\delta\Delta E - \alpha(\mu\mp\nu)^2 + \beta^2\Delta E - \delta\gamma^2 + 2\gamma\beta(\mu\mp\nu)] = 0.$$

Thus, $$2\mathrm{T}_\pm = [(\alpha+\delta) + \frac{1}{\Delta E}\{(\mu\mp\nu)^2+\gamma^2\}] \pm \sqrt{S}$$

where $$S = [(\alpha-\delta)^2 + 4\beta^2]$$
$$\frac{1}{\Delta E}[2(\alpha-\delta)(\gamma^2-(\mu\mp\nu)^2) + 8\gamma\beta(\mu\mp\nu)]$$
$$\frac{1}{(\Delta E)^2}[\gamma^2+(\mu\mp\nu)^2]^2.$$

A little manipulation reduces the first part of S to $B_{\Pi}^2 X^2$ where we have defined

$$X^2 = Y(Y-4) + 4(J+\tfrac{1}{2})^2 \qquad \text{and} \qquad Y = A_{\Pi}/B_{\Pi}$$

We shall now use the fact that ΔE is large: firstly to ignore terms in S dependent on $(1/\Delta E)^2$ and secondly to expand $\sqrt{S}$ as a polynomial ignoring terms higher than first order in $(1/\Delta E)$. This reduces the expression for the roots of the determinant to

$$2T_{\pm} = (\alpha+\delta) + \frac{1}{\Delta E}[(\mu\mp\nu)^2+\gamma^2]$$

$$\pm \frac{B_{\Pi}X + (\alpha-\delta)(\gamma^2-(\mu\mp\nu)^2) + 4\gamma\beta(\mu\mp\nu)}{B_{\Pi}X\Delta E}.$$

We may now proceed to an expression for the Λ-doublet splitting, $\Delta\nu_{ef}$

$$\Delta\nu_{ef} = \frac{(\mu-\nu)^2-(\mu+\nu)^2}{2\Delta E} \mp \frac{-2\mu\nu(\alpha-\delta)+2\gamma\beta(2\nu)}{B_{\Pi}X\Delta E}$$

$$= -(\tfrac{1}{2}p+q)[1\pm(2-Y)/X](J+\tfrac{1}{2}) \mp (2q/X)(J+\tfrac{3}{2})(J+\tfrac{1}{2})(J-\tfrac{1}{2})$$

This is often referred to as the Mulliken and Christy formula where the upper sign refers to F_2 and the lower to F_1 levels and where

$$p = 4\,\frac{\langle {}^2\Pi_{1/2}|B(\mathrm{J}^+\mathrm{L}^- + \mathrm{J}^-\mathrm{L}^+)|{}^2\Sigma^+_{-1/2}\rangle\langle {}^2\Pi_{1/2}|\mathrm{H}_{\mathrm{SO}}|{}^2\Sigma^+_{1/2}\rangle}{\Delta E}$$

$$q = 2\,\frac{\langle {}^2\Pi_{1/2}|B(\mathrm{J}^+\mathrm{L}^- + \mathrm{J}^-\mathrm{L}^+)|{}^2\Sigma^+_{-1/2}\rangle\langle {}^2\Pi_{1/2}|B(\mathrm{L}^+\mathrm{S}^- + \mathrm{L}^-\mathrm{S}^+)|{}^2\Sigma^+_{1/2}\rangle}{\Delta E}$$

In an actual molecule, the $^2\Pi$ vibronic level will be perturbed by many vibronic levels of many $^2\Sigma$ states and so the expressions for p and q become summations. For $^2\Sigma^-$ states, some of the elements in the $\mathbf{P}_+$ and $\mathbf{P}_-$ matrices change sign and there is also a sign change in the linear combinations used in the block-diagonalization. This introduces a factor of $(-1)^k$ into the expressions for p and q. It must be pointed out that there is more than one derivation of these results. We shall now outline a route to these expressions that uses second-order perturbation theory.

Diagonalization of the 4 × 4 matrix for the interaction of $|^2\Pi_{3/2}\rangle$, $|^2\Pi_{1/2}\rangle$, $|^2\Pi_{1/2}\rangle$ and $|^2\Pi_{-3/2}\rangle$, produces two doubly degenerate levels F_1 and F_2. F_1 corresponds to $^2\Pi_{1/2}$ in regular case (a) and to $N = J - \frac{1}{2}$ in Hund's case (b). The wavefunctions obtained from the diagonalization are:

$$|F \pm v\ J\rangle = c_1|^2\Pi_{\pm 1/2}\ v\ J\rangle + c_2|^2\Pi_{\pm 3/2}\ v\ J\rangle$$

where

$$c_1 = \mp\frac{1}{\sqrt{2}}\left[1 \pm \frac{(2-Y)}{X}\right]^{1/2}$$

$$c_2 = \frac{1}{\sqrt{2}}\left[1 \mp \frac{(2-Y)}{X}\right]^{1/2}$$

(the signs in the expressions for the coefficients referring to whether the level is F_1 or F_2).

Using the linear combination

$$P^{\pm} = \frac{1}{\sqrt{2}}(|F + v\ J\rangle \pm |F - v\ J\rangle)$$

$$S^{\pm} = \frac{1}{\sqrt{2}}(|^2\Sigma_{1/2}\ v'\ J\rangle \pm (-1)^k|^2\Sigma_{-1/2}\ v'\ J\rangle)$$

then second-order perturbation theory gives the following expression for the correction to the $^2\Pi$ energy levels

$$E_{\pm}^{(2)} = \sum_{n'\ v'} \frac{\langle P^{\pm}|\mathrm{H}^{(1)}|S^{\pm}\rangle\langle S^{\pm}|\mathrm{H}^{(1)}|P^{\pm}\rangle}{E_{\Pi v} - E_{n'v'}}$$

$$= \sum_{n'\ v'} \frac{[c_1\mu + c_2\gamma \mp (-1)^k c_1\nu]^2}{E_{\Pi v} - E_{n'v'}}.$$

Thus

$$\Delta\nu_{ef} = -4\sum_{n'v'} (-1)^k \frac{c_1^2\mu\nu + c_1c_2\gamma\nu}{E_{\Pi v} - E_{n'v'}}$$

A little algebra reduces this to

$$\Delta\nu_{ef} = -(\tfrac{1}{2}p + q)[1 \pm (2-Y)/X](J+\tfrac{1}{2}) \mp (2q/X)(J+\tfrac{3}{2})(J+\tfrac{1}{2})(J-\tfrac{1}{2}) \tag{3.1}$$

where

$$p_v = 4 \sum_{n' v'} \frac{(-1)^k \langle {}^2\Pi \ v | \mathrm{H_{SO}} | n'v' \rangle \langle n'v' | B(\mathrm{L^+S^-} + \mathrm{L^-S^+}) | {}^2\Pi \ v \rangle}{E_{\Pi v} - E_{n'v'}}$$

$$q_v = 2 \sum_{n' v'} \frac{(-1)^k | \langle {}^2\Pi \ v | B(\mathrm{L^+S^-} + \mathrm{L^-S^+}) | n'v' \rangle |^2}{E_{\Pi v} - E_{n'v'}}$$

(3.2)

The summations extend over all ${}^2\Sigma$ vibronic levels $|n'v'\rangle$; k is even for ${}^2\Sigma^+$ states and is odd for ${}^2\Sigma^-$ states; the upper sign relates to F_2 levels and the lower sign to F_1:

$$X^2 = Y(Y-4) + 4(J+\tfrac{1}{2})^2 \qquad \text{and} \qquad Y = A_\Pi / B_\Pi$$

(3.3)

Consideration of the form of ($\mathrm{H_{rot}} + \mathrm{H_{SO}}$) shows that interactions of ${}^2\Pi$ are possible not only with ${}^2\Sigma$ states but also with ${}^4\Sigma$ and ${}^2\Delta$ states. These latter effects are usually small. We shall see examples later of other states having significant contributions.

In the case of $J = \frac{1}{2}$, X simplifies to $(Y - 2)$ and so:

$$\Delta\nu = -(p + 2q)$$

(3.4)

The formulae we have obtained for the Λ-doubling parameters explicitly deal with independent ${}^2\Sigma$ -${}^2\Pi$ interactions. We can envisage that problems might arise if there were several states of the *same* symmetry which were sufficiently close to perturb each other strongly.

3.2.3. *Calculation of off-diagonal matrix elements of* $\mathrm{H_{SO}}$ *and* $\mathrm{L^+}$

The calculation of off-diagonal matrix elements of the spin–orbit coupling operator is greatly simplified if all the determinants are constructed from the same set of molecular orbitals. This is referred to as the invariant orbital approximation and is useful because Slater's rules apply; in particular matrix elements between determinants which differ by more than two spin-orbitals vanish. If separate SCF procedures are carried out on each state then orbitals cease to be orthogonal between states and the rules cease to apply. Use of a common basis set still leads to some simplification of the integrals.

It has been shown by Löwdin that if determinants ψ_p and ψ_q are constructed from the same atomic basis set $\{\chi_i\}$ and if Ω is an operator which can be expressed as the sum of one-electron operators ω then:

$$\langle\psi_p|\Omega|\psi_q\rangle = \sum_{ijab} \{c_{ib}c_{ja}\ \Delta_{ij}(S_{pq})\ \langle\chi_a|\omega|\chi_b\rangle$$

where i and j label molecular spin-orbitals, a and b label atomic basis functions, c_{ib} is the coefficient of basis function b in molecular spin-orbital i, and $\mathbf{\Delta}$ is the matrix of cofactors of S_{pq}, the matrix of overlap between molecular spin-orbitals.

Hall has written a program which calculates one-centre one-electron and one-centre two-electron integrals between Slater determinants which have non-orthogonal orbitals but which have a common basis set: attention is restricted to pairs of determinants which differ in only one spin-orbital that is $\bar{\pi}^+$ in the first state and σ in the other. The matrix of cofactors is obtained by the method of Prosser and Hagstrom which involves diagonalization of the overlap matrix by biorthogonalization. The two-electron integrals are evaluated using the transformation equation:

$$\langle\psi_p|\mathrm{H_{SO}}|\psi_q\rangle = \sum_{abcd} (\rho_{ac}\rho_{bd} - \rho_{ad}\rho_{bc})\langle\chi_a\chi_b|\omega|\chi_c\chi_d\rangle$$

where

$$\rho_{ab} = \sum_{ij} \{c_{ib}c_{ja}\ \Delta_{ij}(D_{pq})\}.$$

The matrix D_{pq} is identical to S_{pq} where a is a σ orbital and b is a $\bar{\pi}^+$ orbital. Otherwise, it is obtained from S_{pq} by deleting the odd $\bar{\pi}^+$ and σ spin-orbitals. More details concerning Hall's program may be found on the microfiche and in the further reading.

The operator L^+ may be expanded as a sum of one-electron operators (l_i^+) and so there is a close resemblance to the calculation of the one-electron part of matrix elements of $\mathrm{H_{SO}}$. For matrix elements involving STOs centred on the same nucleus A, the operator l^+ may be expanded as:

$$\mathrm{l}^+ = \mathrm{l}_A^+ - R_m\nabla^+$$

where l_A^+ is the shift operator for one-electron orbital angular momentum about centre A and R_m is the distance between A and the centre of mass; notice that L^+, unlike $\mathrm{H_{SO}}$, varies with different reduced mass so that OH and OD, for example, will have identical matrix elements of H_{SO} but different matrix elements of L^+.

Using the fact that STOs centred on A are eigenfunctions of l_A^2 and using the formula given by Edmonds for the effect of the gradient operator on a spherical harmonic function, it is possible to deduce an expression for the effect of l^+ on a σ-type STO. Using this expression, one-centre integrals of the form $\langle\pi^+|l^+|\sigma\rangle$ become analytical.

Integrals involving STOs on different centres may be evaluated by expanding the operator l^+ as

$$l^+ = \frac{m_A}{m_A+m_B} l_A^+ + \frac{m_B}{m_A+m_B} l_B^+$$

where m_A and m_B are nuclear masses. Thus

$$\langle\chi_A|l^+|\chi_B\rangle = \frac{m_A}{m_A+m_B}\langle\chi_B|l_A^-|\chi_A\rangle + \frac{m_B}{m_A+m_B}\langle\chi_A|l_B^+|\chi_B\rangle$$

These integrals over STOs (χ_i) resemble overlap integrals; the angular eigenfunctions are orthonormal and so all that remains is the radial integration. This must be performed numerically and may be readily carried out using, for example, crossed Gaussian–Legendre quadrature.

3.2.4. *Examples: BeH, CH, OH, and NO*

Estimates of the Λ-doubling constants in the ground states of these molecules were made as early as 1931 by Mulliken and Christy. Certain rather gross assumptions were used including the 'pure precession approximation' and the assumption that diagonal and off-diagonal matrix elements are equal.

In the limit of pure-precession, matrix elements arise from a single electron with orbital angular momentum l, whose projection onto the molecular axis is zero in one state and unity in the other. In this case, the matrix elements of L^+ become $\sqrt{(l(l+1))} = \sqrt{2}$. Factorizing the vibronic integrals gives:

$$\langle {}^2\Pi \;\; v|BL^+|{}^2\Sigma^+ \;\; v'\rangle \sim \sqrt{2}B_\Pi$$

$$\langle {}^2\Pi \;\; v|H_{SO}|{}^2\Sigma^+ \;\; v'\rangle \sim A_\Pi/\sqrt{2}$$

where it has been assumed that the vibrational wavefunctions of the different states form an orthonormal set so that the overlap integral is unity for the 0–0 perturbation, and zero otherwise. The electronic matrix elements of H_{SO} (which is still often written as $\frac{1}{2}AL^+$ even though it cannot be literally factorized in this way) have been replaced by $A_\pi/\sqrt{2}$; the one-electron matrix elements are of the correct form but the radial integration is between different orbitals and the approximation is invalid for the two-electron terms.

For CH, interactions of $X^2\Pi_r$ with both $B^2\Sigma^-$ and $C^2\Sigma^+$ states were included. Mulliken and Christy assumed that pure precession was appropriate for the individual angular momenta l and reduced the expressions for p and q to

$$p_0 = 4A_\Pi B_0 \left\{ \frac{1}{\nu(X^2\Pi, C^2\Sigma^+)} - \frac{1}{\nu(X^2\Pi, B^2\Sigma^-)} \right\}$$

$$q_0 = 4B_0^2 \left\{ \frac{1}{\nu(X^2\Pi, C^2\Sigma^+)} - \frac{1}{\nu(X^2\Pi, B^2\Sigma^-)} \right\}.$$

Use of experimental values for the spin–orbit coupling constants, the rotational constants, and the energy differences gave reasonable results for a number of molecules. In a few cases, the differences between experimental and estimated values of p and q were considerable.

Better values may be obtained by carrying out an *ab initio* study. Hinkley calculated electronic matrix elements from Hartree–Fock wavefunctions and used rather more realistic vibrational terms. However, no account was taken of the variation of the electronic integrals with internuclear distance – this is referred to as the R-centroid approximation. Provided these matrix elements are constant or vary linearly over the range of internuclear distance for which vibrational overlap is significant, this approximation gives very good results. Note that the vibrational analysis is sometimes carried out at R_e rather than at the R-centroid of the potential curves.

OH. Table 3.1 gives various contributions to the matrix element $\langle X^2\Pi | H_{SO} | A^2\Sigma^+ \rangle$ for OH. An atomic population analysis confirms that the change occurring in going from $X^2\Pi$ to $A^2\Sigma^+$ is essentially a $2p_\sigma$ orbital on oxygen going to a $2p_\pi$ orbital on oxygen. Indeed, it can be seen that the contribution from the H atom to the off-diagonal matrix element of H_{SO} is very small. The corresponding matrix elements of L^+ are given in Table 3.2 – the total value is found to be close to $\sqrt{2}$. All of these matrix elements were calculated using the invariant orbital approximation.

Table 3.1
Contributions to $\langle X^2\Pi | \mathrm{H_{SO}} | A^2\Sigma^+ \rangle$ *in OH (all values in* cm^{-1}*)*

		O atom	H atom
One-electron contribution		−248.96	−0.07
Two-electron contribution			
Closed shell	Direct	46.74	0.00
	Exchange	−36.84	0.00
Open shell	Direct	10.65	0.00
	Exchange	3.22	0.00
Total		−157.95	−0.07

Table 3.2
Contributions to $\langle X^2\Pi | \mathrm{L}^+ | A^2\Sigma^+ \rangle$ *in OH*

	OH	O atom	H atom
l_i^+	1.1305	1.1206	−0.0003
$\sqrt{2}R_i\nabla_i^+$	0.2259	0.0155	0.0141
Total	1.3564	1.1361	0.0138

The vibrational eigenfunctions are determined in numerical form on a grid by solution of the radial Schrödinger equation: a Numerov procedure. The differential equation is replaced by a difference equation of the same order and an initial guessed eigenvalue is used to generate a solution which does not satisfy the boundary conditions. The deviation from these conditions is used to generate a better eigenvalue and the process is repeated until the boundary conditions are satisfied within a reasonable tolerance. Care must be taken to ensure that neither limit of integration truncates a tail of the wavefunction and that there are sufficient points per node. Simpson's rule integration may then be used to produce the required vibrational matrix elements. It is found that Morse curves and RKR curves give almost identical results for OH. Consideration of the expressions for p and q shows that the vibrational matrix elements of interest are $\langle v''|v'\rangle\langle v''|\mathrm{B}|v'\rangle$ and $|\langle v''|\mathrm{B}|v'\rangle|^2$. These are tabulated below (Table 3.3) for the $v'' = 0$ and $v'' = 1$ levels of $X^2\Pi$ of OH and the $v' = 0$ to $v' = 12$ levels of $A^2\Sigma^+$.

Table 3.3
Vibrational matrix elements for OH

	$\langle v''\|v'\rangle\langle v''\|\mathrm{B}\|v'\rangle$		$\|\langle v''\|\mathrm{B}\|v'\rangle\|^2$	
v'	$v''=0$	$v''=1$	$v''=0$	$v''=1$
0	15.95495	0.89625	282.8541	8.3447
1	2.22797	11.55229	54.9429	196.0602
2	0.28929	4.12761	9.8539	92.5905
3	0.04463	0.94204	2.0306	27.9112
4	0.00875	0.21415	0.5062	8.2059
5	0.00230	0.05549	0.1524	2.6429
6	0.00075	0.01712	0.0543	0.9626
7	0.00030	0.00628	0.0222	0.3959
8	0.00013	0.00267	0.0102	0.1808
9	0.00007	0.00126	0.0050	0.0890
10	0.00003	0.00062	0.0025	0.0449
11	0.00002	0.00028	0.0012	0.0205
12	0.00000	0.00005	0.0002	0.0039

We may make a number of observations from these figures. The values are swamped by the term with $v'' = v'$ so that we can understand why reasonable results are given by the assumption that vibrational wavefunctions of different states form an orthonormal set. The values show that convergence is not a serious problem and that very few vibrational levels need to be considered. Further, the magnitude of the $v' = 12$ matrix elements suggests that higher bound levels and continuum levels may be safely neglected.
Consideration of just the $A^2\Sigma^+$ state $(1\sigma^2 2\sigma^2 3\sigma 1\pi^4)$ gives values of p_0 and q_0 for the $X^2\Pi_i$ state of OH $(1\sigma^2 2\sigma^2 3\sigma^2 1\pi^3)$ that are remarkably close to experiment (Table 3.4).

Table 3.4
p_0 and q_0 for OH

	p_0/cm^{-1}	q_0/cm^{-1}
Mulliken and Christy	0.311	−0.0417
Hinkley	0.242	−0.0391
Terrestrial experiment	0.2357	−0.0388

BeH. The ground state of BeH is $X^2\Sigma^+$ $(1\sigma^2 2\sigma^2 3\sigma)$ and the first excited state is $A^2\Pi_r$ $(1\sigma^2 2\sigma^2 1\pi)$. In this case the transition involved in going from the A state to the X state is $2p_\sigma$ on Be going to $2p_\pi$ on Be. The calculations (see Table 3.5) give fair agreement with experiment and after these successes for OH and BeH, we would expect good results for CH. We shall soon see that the values obtained for this molecule are rather disappointing.

Table 3.5

p_0 and q_0 for BeH

	p_0/cm^{-1}	q_0/cm^{-1}
Mulliken and Christy	0.00405	0.0212
Hinkley	0.00142	0.0134
Terrestrial experiment	—	0.0142

CH. The ground state of CH is $X^2\Pi_r$ $(1\sigma^2 2\sigma^2 3\sigma^2 1\pi)$ and the first $^2\Sigma$ states encountered are $B^2\Sigma^-$ and $C^2\Sigma^+$ which are both derived from the electronic configuration $1\sigma^2 2\sigma^2 3\sigma 1\pi^2$. In the united atom limit the 3σ and 1π orbitals are degenerate: if pure precession were to hold for l then the molecular integral $\langle 1\pi^+ | l^+ | 3\sigma \rangle$ would have the value $\sqrt{2}$ and all other similar integrals would be zero. From this argument, only $B^2\Sigma^-$ and $C^2\Sigma^+$ need be included in the perturbation treatment. This truncation of the summations in the expressions for p and q is frequently referred to as the 'pure precession hypothesis'. Clearly, we have met this term in a slightly different context and confusion is easy! It is useful to examine the matrix element $\langle X^2\Pi | \mathrm{L}^+ | C^2\Sigma^+ \rangle$ in the invariant orbital approximation. The Hartree–Fock wavefunctions for the Σ states are actually linear combinations of Slater determinants and this introduces numerical factors neglected by Mulliken and Christy,

$$\begin{aligned}\langle X^2\Pi|\mathrm{L}^+|C^2\Sigma^+\rangle &= \frac{1}{\sqrt{2}}\langle 3\sigma 3\bar{\sigma}\, 1\pi^+|\mathrm{L}^+|3\sigma\, 1\pi^+ 1\bar{\pi}^-\rangle \\ &\quad - \frac{1}{\sqrt{2}}\langle 3\sigma 3\bar{\sigma}\, 1\pi^+|\mathrm{L}^+|3\sigma\, 1\bar{\pi}^+ 1\pi^-\rangle \\ &= -\frac{1}{\sqrt{2}}\langle 3\sigma|\mathrm{L}^+|1\pi^-\rangle.\end{aligned}$$

Assuming pure precession of l, the matrix element reduces to the value -1. Hinkley used Morse curves for the $^2\Sigma$ states and obtained the values in Table 3.6. The Λ-doubling predicted for the rotational ground state is too small by about 20 per cent despite the much better agreement found in the similar calculations on BeH and OH.

Table 3.6

$(\frac{1}{2}p_0 + q_0)$ *and* q_0 *for CH*

	$(\frac{1}{2}p_0 + q_0)/\mathrm{cm}^{-1}$	q_0/cm^{-1}
Mulliken and Christy	0.0108	0.0054
corrected	0.0611	0.0334
Hinkley	0.0447	0.0254
Terrestrial experiment	0.0563	0.038

We shall see later why this is so and will discover how excellent agreement may be found with astronomical measurement. The idea that only one or two states need be considered ('pure precession') has been central to the argument so far and it is important to examine the validity of this idea. Hund's case (a) molecular wavefunctions are not eigenfunctions of L^2 since only L_z commutes with the Hamiltonian: L is not a good quantum number. Indeed, this is the reason why we must calculate matrix elements of L^+ whereas other operators such as S^- can be treated analytically.

Colbourn and Wayne have considered the molecules NH, NF, PH, and PF all of which have $^3\Sigma^-$ ground states and have written

$$\langle L^2\rangle = \langle L_z^2\rangle - 2\sum_{^3\Pi} \langle ^3\Sigma^-|L^-|^3\Pi\rangle\langle ^3\Pi|L^+|^3\Sigma^-\rangle$$

$$= 2\sum_{^3\Pi} |\langle ^3\Sigma^-|L^-|^3\Pi\rangle|^2$$

where the summation is over $^3\Pi$ states. It is found that large values of $\langle L^2\rangle$ arise for non-hydride molecules indicating that many states contribute significantly to the summation. In particular, the value for PF is two orders of magnitude greater than that for PH (see Table 3.7). The pure precession hypothesis breaks down for many non-hydride molecules but remains a useful concept for molecules which do not deviate markedly from the united-atom limit.

Table 3.7
Values of $\langle L^2 \rangle$

Molecule	R/bohr	$\langle L^2 \rangle$	$\langle ^3\Sigma^- \vert L^- \vert ^3\Pi \rangle$
NH	1.961	4.32	0.907
NF	2.490	272.77	0.986
PH	2.708	8.40	0.867
PF	3.005	909.61	0.814

NO. It is interesting to examine the case of NO for which there are a very large number of low-lying $^2\Sigma$ states. In principle, the Λ-doubling in the $^2\Pi$ ground state $(1\sigma^2 2\sigma^2 3\sigma^2 4\sigma^2 5\sigma^2 1\pi^4 2\pi)$ is mainly due to the $G^2\Sigma^-$ state $(1\sigma^2 2\sigma^2 3\sigma^2 4\sigma^2 5\sigma 1\pi^4 2\pi^2)$ and restriction of the analysis to just this Σ state gives the poor values in Table 3.8. It is tempting to explain this result as the breakdown of pure precession.

Table 3.8
p_0 and q_0 for NO

	p_0/cm^{-1}	q_0/cm^{-1}
Mulliken and Christy	0.0115	0.0014
Hinkley	0.0039	0.00018
Experiment	0.0117	0.000077

The accuracy of the calculation. It is rather unrealistic to assign the poor results for CH to the same cause suggested for NO. It has been shown experimentally that the $B^2\Sigma^-$ and $C^2\Sigma^+$ states of CH both have potential maxima and thus are very poorly represented by Morse curves. Hammersley calculated vibrational wavefunctions from extended CI curves assuming the hypothetical case of $J = 0$ and neglecting the effects of predissociation in excited levels. These when combined with Hinkley's electronic matrix elements produced still poorer Λ-doubling constants, labelled (a) in Table 3.9.

The invariant orbital approximation is capable of producing large errors and so Hammersley carried out a separate SCF procedure for each state. The Λ-doublet splitting then obtained, labelled (b), was restored to Hinkley's values – further improvements were obviously necessary. It was configuration interaction which led to a large

Table 3.9
Λ-doubling constants for CH

	$(\frac{1}{2}p + q)/\text{cm}^{-1}$	q/cm^{-1}	Splitting (MHz)
Hinkley	0.0447	0.0254	2680
Hammersley (a)	0.0430	0.0245	2576
(b)	0.0447	0.0274	2680
(c)	0.0563	0.0386	3374
(d)	0.0552	0.0379	3311
Terrestrial experiment	0.0563	0.038	3374 ± 20
Astronomical experiment	0.0556	–	3335.47 ± 0.01

change in the values of p and q. Six configuration state functions were included for the $X^2\Pi$ and $B^2\Sigma^-$ states while eight were used for the $C^2\Sigma^+$ state. The results, labelled (c), were found to be in excellent agreement with terrestrial experiment. The CH molecule was at that time unknown in the interstellar medium despite much searching close to the frequency estimated from high resolution optical spectra. It was felt that some further improvement might still be possible and the importance of the R-centroid approximation was examined. Electronic matrix elements were calculated at a number of different internuclear distances using CI wavefunctions and were integrated over the vibrational wavefunctions. During the course of this work, observation of CH in interstellar space was reported. The results of this last improvement, labelled (d), are particularly pleasing since they show that the Λ-doubling is predicted more accurately by *ab initio* calculation than by terrestrial experiment.

These remarkably good values for CH pose a very serious threat not only to the experimentalist but also to the theoretician. The question to be asked is whether the large changes that occurred when better potential curves and improved electronic wavefunctions were used would be repeated in other molecules. Hinkley's values for OH did not approach the order of accuracy demanded for the accurate prediction of astronomical frequencies but were remarkably close to terrestrial values. Unless 'improvement' of the wavefunctions for OH produced better Λ-doubling constants, the values for CH would be little more than fortuitous. Cooper has calculated Λ-doubling constants for OH using separate SCF procedures on the $X^2\Pi$ and $A^2\Sigma^+$ states. The wavefunctions were improved using 15Π configuration state functions (CSFs) and 11Σ CSFs. Table 3.10 shows that the

Table 3.10

Λ-doubling constants for OH

	p/cm^{-1}	q/cm^{-1}	Splitting (MHz)
Mulliken and Christy	0.311	−0.0417	
Hinkley	0.242	−0.0391	
Cooper, no CI	0.2308	−0.0381	
with CI	0.2378	−0.0398	1666
Terrestrial experiment	0.2357	−0.0388	1666

changes due to dropping the invariant orbital approximation and to including more CI move the values in opposite directions and, to a certain extent, the two effects tend to cancel. Agreement with terrestrial experiment would appear to be excellent and our fears would appear to have been groundless.

Fig. 3.2 indicates the observed astronomical transitions with epicentres separated ($\Delta\nu$) by 1666 MHz. There is excellent agreement of

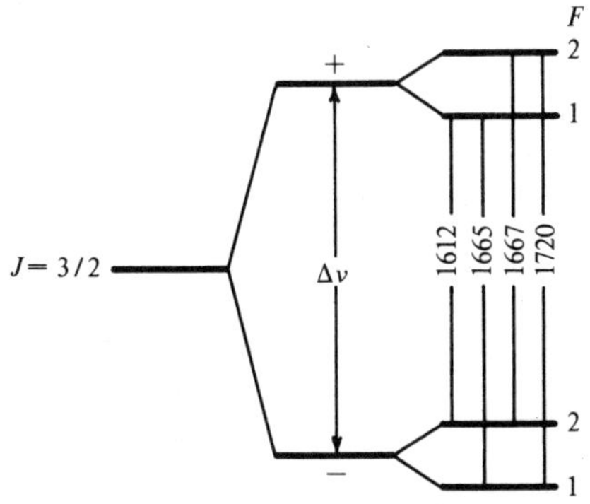

Fig. 3.2. Λ-doubling ($\Delta\nu$) in the lowest rotational level of OH ($^2\Pi_{3/2}$; $J = 3/2$) showing the observed radioastronomical transitions in MHz (not to scale).

the *ab initio* value with astronomical observations. Table 3.11 gives a few of the more recent terrestrial values of p and q. The variation is quite marked and we may feel justified in asserting that *ab initio* values for Λ-doublet splittings may be more reliable than estimates from terrestrially measured electronic spectra particularly when very high-resolution spectra are unobtainable for one reason or another. For $J = \frac{1}{2}$ of $^2\Pi_{1/2}$, the separation of epicentres of the observed astronomical lines is 4730 MHz for which the calculated values of

Table 3.11
Experimental values of p *and* q *for OH*

	p_0/cm^{-1}	q_0/cm^{-1}
Poynter and Beaudet	0.2349	−0.0388
Moore and Richards	0.246	−0.0384
Meerts and Dymanus	0.2357	−0.0388
Destombes *et al.*	0.2348	−0.0387

p and q predict 4795 MHz; $J = 5/2$ of $^2\Pi_{3/2}$ has a terrestrially measured separation of 6033 MHz to be compared with 5946 MHz; the even more highly excited level $J = 5/2$ of $^2\Pi_{1/2}$ has an epicentric separation of 8153 MHz to be compared with 8444 MHz. In the manifold excited rotational states, errors are magnified by multiplication of the constants by terms up to cubic in the rotational quantum number. However, agreement is good even in the most highly excited of these levels.

Possible errors remaining in the calculations for OH include the R-centroid approximation, the neglect of other Σ states and the neglect of third-order perturbation terms. The validity of the R-centroid approximation may be easily demonstrated but it must be stressed that the position of the R-centroid would change with increased vibrational quantum number. The effects of $B^2\Sigma^+$ have been calculated to be less than those of $A^2\Sigma^+$ by a factor of 10^9 due to much reduced vibrational overlap. We may thus safely ignore states such as $B^2\Sigma^+$ and $C^2\Sigma^+$. Third- and fourth-order perturbation terms have been considered experimentally and have been used to produce better agreement between terrestrial experiment and astronomical observation: all of these terms are very small. It is also possible to introduce terms due to deviation from Hund's case (a) or case (b) behaviour – these are also very small except for very high values of J.

3.2.5. *The inclusion of second-row atoms: SiH*

The cosmic abundance of Si suggests the presence of SiH in interstellar clouds but this molecule has not yet been observed. In view of the results obtained for other molecules, Wilson carried out calculations of p and q for this molecule. Within the pure-precession hypothesis the only states important in the summations for p and q in the $X^2\Pi_r$ state ($1\sigma^2 2\sigma^2 3\sigma^2 4\sigma^2 5\sigma^2 1\pi^4 2\pi$) are $B^2\Sigma^+$ and an unobserved, unbound $^2\Sigma^-$ state both of which arise from the configuration

$1\sigma^2 2\sigma^2 3\sigma^2 4\sigma^2 5\sigma^1 1\pi^4 2\pi^2$. Separate SCF procedures were carried out for each state and electronic matrix elements were evaluated over a range of internuclear distance using CI wavefunctions. The first problem that arises is the uncertainty in the energy separation between the $X^2\Pi$ and $^2\Sigma$ states: for the $B^2\Sigma^+$ state, experimental values may be used but for the unbound $^2\Sigma^-$ state, a calculated value must be used. The second problem arises from the fact that the $^2\Sigma^-$ state has no minimum so that a continuum of vibrational levels must be considered. 'Pseudo-continuum' wavefunctions were calculated using a procedure known as box normalization: the system was treated as a bound state by placing a very steep potential barrier at large internuclear distance. It was found that moving the wall from 25 Å to 30 Å caused a change of less than 2 per cent. This procedure has been fairly widely used: calculations on the predissociation of O_2 agree well with results obtained by other methods. Unfortunately, a small percentage error in matrix elements involving these continuum vibrational levels leads to a large error (in MHz) in the frequency prediction for the Λ-doubling. The value obtained by Wilson for the splitting was 3168 MHz as compared to a recent terrestrial measurement of 2968 MHz. The three most obvious sources of error are the quality of the vibrational wavefunctions for $^2\Sigma^-$, the uncertainty in term values, and the neglect of core polarization.

If a continuum wavefunction with energy E and rotational quantum number J is denoted by $\psi_{E,J}(R)$ then the appropriate radial Schrödinger equation is

$$\{d^2/dR^2+2\mu(E-V(R))/\hbar^2-J(J+1)/R^2\}\psi_{E,J}(R)=0$$

with boundary conditions

$$\psi_{E,J}(0)=0$$

$$\lim_{R\to\infty} \psi_{E,J}(R)=\left[\frac{2\mu}{\pi k h^2}\right]^{1/2} \sin(kR-J\pi/2+\eta_J(E))$$

where $\eta_J(E)$ is the phase shift and $k^2=2\mu E/\hbar^2$.

The boundary condition for large R may be easily imposed if the asymptotic amplitude, $A_\infty(R)$, of the wavefunction is known. Use of the first-order WKB approximation gives an expression for this amplitude:

$$A_{\infty}(R)^4 = A(R)^4 \left\{ \frac{E - V(R) - \hbar^2 J(J+1)/2\mu R^2}{E} \right\}$$

Successive maxima may be fitted to this expression so as to find the asymptotic amplitude. This procedure leads to a much smaller range of integration than would integration until the wavefunction was sinusoidal within a given tolerance. For continuum levels, the summations over vibrational levels in the expressions for p and q must be replaced by integrals with respect to energy. Cooper has used very large CI expansions (several hundred CSFs) to produce a better estimate of the energy separation between $X^2\Pi$ and $^2\Sigma^-$ and has calculated continuum wavefunctions for $^2\Sigma^-$. The calculated Λ-doubling constants have been collected in Table 3.12.

Table 3.12

Calculated Λ-doubling constants for SiH

	p_0/cm^{-1}	q_0/cm^{-1}	Splitting (MHz)
Wilson	0.053927	0.012727	3168
Cooper $B^2\Sigma^+$	0.028673	−0.001185	
$^2\Sigma^-$	0.053363	0.010200	
Total	0.08204	0.09015	3000

Published terrestrial values differ not only with the resolution of the observed electronic spectrum but also with the method of analysis: some workers use a term value approach whereas others prefer a direct fit procedure. Various predictions of the Λ-doublet splitting in the $v'' = 0$, $J = \frac{1}{2}$ level f the $X^2\Pi$ state of SiH are collected in Table 3.13. The *ab initio* value is closer to the most recent terrestrial value than are many of the earlier predictions. The true accuracy of the prediction of 3000 MHz will not be known until SiH is observed in the interstellar medium. Consideration of the diagonal matrix elements of H_{SO} suggests that the one-centre approximation leads to an error of only about 1/4 per cent but also suggests that core polarization is important. If core polarization is visualized as polarization of the $1\pi^4$ closed shell by the $2\pi^1$ electron then the effect would be expected to be small in the $^2\Sigma$ states where the two

Table 3.13
Terrestrial predictions of the Λ-splitting in SiH

	Splitting (MHz)	Estimated error (MHz)
Douglas, Elliott	2940	
Klynning, Lindgren	2942	±30
Freedman, Irwin	2932	±18
Klynning *et al.* (1979)	2968	±6

2π electrons have opposing contributions. This argument coupled with the observation that the calculated Λ-doublet splitting is within about 1 per cent of terrestrial experiment suggests that core polarization does not pose serious problems. Neglect of terms higher than second-order and neglect of other states leads to smaller errors than the uncertainty in the energy separation between $X^2\Pi$ and $^2\Sigma^-$.

3.2.6. *HF⁺ and NH⁺*

HF^+ is isoelectronic with OH but there are interesting differences: it is necessary to allow the summations over vibrational levels of the $A^2\Sigma^+$ state to extend over continuum levels as well as over bound levels. Wilson treated the bound levels and the continuum levels independently and obtained parameters in reasonable agreement with experiment. In higher vibrational levels of the $X^2\Pi_i$ state, the contribution of $A^2\Sigma^+$ continuum levels was greater than that of the bound levels.

Hutson and Cooper have repeated the HF^+ calculations using a method which directly calculates the vibrational parts of p and q without explicit summation. This ensures that contributions from the top of the ladder of the bound levels or from the bottom of the continuum are not excluded. The expressions for p and q reduce to

$$p = 4\sum_{n'} (-1)^k \langle {}^2\Pi_v | \mathrm{H_{SO}} | n'\omega' \rangle$$

$$q = 2\sum_{n'} (-1)^k \langle {}^2\Pi_v | B(\mathrm{L^+S^-} + \mathrm{L^-S^+}) | n'\omega' \rangle$$

where the summation is over electronic states and the electronic part of the ket $|n'\omega'\rangle$ is just the appropriate electronic wavefunction. $|n'\omega'\rangle$ is the first-order correction to the state due to the perturbation; it is not a normalized function and ω' is merely a label and not a quantum number. The vibrational part is obtained by solution of the inhomogeneous differential equation

$$\left[\frac{-\hbar^2}{2\mu}\frac{d^2}{dR^2} + U_{n'}(R) - E_{\Pi v}\right] \chi_{n'\omega'}(R) = B(R)H'_{n'}(R)\chi_{\Pi v}(R)$$

where $U_{n'}(R)$ is the effective potential curve for state n' and $\chi_{\Pi v}(R)$ is the vibrational part of the wavefunction for the perturbed state. The quantity $H'_{n'}$ is the electronic part of the matrix element of the shift operators:

$$H'_{n'}(R) = \int \phi^*_{n'}(L^+S^- + L^-S^+)\phi_\Pi d\tau_e$$

where τ_e signifies that the integral is over all electronic coordinates but not over the internuclear distance R.

This method has been rigorously tested for a number of systems. In the case of HF^+, results are given that are believed to be more reliable than those from electronic spectroscopy. Observed and calculated Λ-doubling parameters are given in Table 3.14.

Table 3.14
Calculated and observed Λ-doubling parameters in the $X^2\Pi_i$ state of ${}^1H^{19}F^+$

	p_v/cm^{-1}		q_v/cm^{-1}	
v	Calculated	Observed	Calculated	Observed
0	0.589	0.64 ± 0.06	−0.0402	−0.046 ± 0.008
1	0.580	0.63 ± 0.08	−0.0382	−0.046 ± 0.008
2	0.573	0.66 ± 0.08	−0.0363	−0.049 ± 0.010

Yet another example of the unreliability of arguments assuming similar behaviour of isoelectronic species is provided by NH^+ which might be expected to resemble CH. The first excited state is a very low-lying ${}^4\Sigma^-$ state which causes strong perturbations in the optical spectrum. The $X^2\Pi - a^4\Sigma^-$ interaction may be treated by setting up the appropriate perturbation matrix and block-diagonalizing it by using suitable parity eigenfunctions. The states are then allowed to interact with the new levels obtained from the diagonalization.

Wilson used this procedure to find the Λ-doubling in the lowest vibrational level of $X^2\Pi$, $J = \frac{1}{2}$. He obtained 13 625 MHz as opposed to an experimental value of 13 520 MHz. This agreement is excellent. The neglect of the $a^4\Sigma^-$ state by use of the Mulliken and Christy formula gives 6839 MHz.

3.2.7. *BeF*

We have already mentioned the case of BeF whose first excited state is $A^2\Pi_r$ but has values of p and q which have opposite signs. The results of Walker are shown in Table 3.15; a number of approximations were used to obtain a positive value for p, as required. Apart from the HF matrix elements, the largest contribution was from the inverted $^2\Pi_i$ state (---$5\sigma^2 1\pi^3$) but its contribution was very small compared with that from the regular state (---$1\pi^4 2\pi$). Very little CI was used and a number of states were neglected but Walker was able to reproduce the correct signs for the Λ-doubling parameter. Prosser has repeated the calculations including other Σ states and more CI. The results are disappointing and this problem would again appear to be unsolved.

Table 3.15
Calculated and observed Λ-doubling parameters in the $A^2\Pi$ state of BeF

		$q_0 \times 10^4/\text{cm}^{-1}$	$p_0 \times 10^4/\text{cm}^{-1}$
Experiment:	Walker & Barrow	−3.0 ± 0.5	−7.6 ± 1.0
	Gurvich & Novikov	−2.96	7.98
Calculation:	Walker & Richards	−3.26	0.10
	Prosser & Richards	−1.51	−1.54

The order of states is $X^2\Sigma^+$, $A^2\Pi_r$, $B^2\Sigma^+$ and so $X^2\Sigma^+$ has positive contributions to p and q while $B^2\Sigma^+$ has negative contributions to p and q in the $A^2\Pi_r$ state. The problem is further complicated by the avoided crossing of $B^2\Sigma^+$ and $C^2\Sigma^+$. The final values of either p or q can be of either sign depending on the relative magnitudes of the different competing effects. The *ab initio* values are thus extremely sensitive to the quality of the wavefunctions and to the number of states considered.

The Mulliken and Christy expression for the Λ-doubling adequately deals with individual Σ–Π interactions but takes no account of indirect contributions via Σ–Σ interactions. This may be one reason for the poor results for BeF, and possibly even for NO.

3.3. Spin-doubling in $^2\Sigma$ states and g values in e.s.r.

Part of the Breit Hamiltonian causes a direct interaction between the spin and rotational angular momenta. However, this effect is usually small compared with an indirect interaction due to H_{SO}. Orbital angular momentum is mixed with the spin angular momentum through off-diagonal matrix elements of the spin–orbit coupling operator; this in turn leads to a coupling between the spin and rotational angular momenta.

The spin-doubling may be described using an effective spin–rotation Hamiltonian:

$$H_{sr} = \gamma \mathbf{N}\cdot\mathbf{S} \tag{3.5}$$

where **N** is the rotational angular momentum of the nuclear framework, **S** is the total electron spin angular momentum, and γ is the spin–rotation coupling constant. The first-order contribution to the spin–rotation coupling constant in a given $^2\Sigma$ vibrational level is:

$$\gamma_v^{(1)} = \frac{-ge\hbar^2}{2m_e c^2\langle I\rangle}\langle ^2\Sigma \;\; v|\sum_K\sum_i \frac{Z_k\cos\theta_{Ki}}{r_{iK}^2}|^2\Sigma \;\; v\rangle \tag{3.6}$$

where the summation extends over all nuclei, K, and over all unpaired electrons, i, and where $\langle I\rangle$ is the vibrational expectation value of the moment of inertia. The second-order contribution may be derived by considering the interaction between a $^2\Sigma$ state and a $^2\Pi$ state using second-order perturbation theory. The expression obtained closely resembles that derived above for p except that the summation is now over all $^2\Pi$ states:

$$\gamma_v^{(2)} = 4\sum_{n' \; v'} \frac{\langle ^2\Sigma \;\; v|BL^+|n'v'\rangle\langle n'v'|H_{SO}|^2\Sigma \;\; v\rangle}{E_{n'v'} - E_{\Sigma v}}. \tag{3.7}$$

There is high correlation between the spin-rotation coupling constant γ and the centrifugal distortion A_D in the spin-orbit coupling constant. Separate determinations of γ and A_D have been obtained only in special cases when high-quality spectra are available for isotopic molecules. The description of Zeeman spectra is strongly affected by the choice of γ or A_D – this leads to different values of fitted g values in e.s.r. *Ab initio* calculations of γ (or of g values) are obviously of some importance to the experimentalist.

Hall has used approximate matrix elements to calculate $\gamma^{(2)}$ for AlO and has obtained a value of 0.0029 cm^{-1} as opposed to a pure precession value of 0.005 cm^{-1} and an experimental value of 0.008 ± 0.006 cm^{-1}.

In the case of HF^+, rather better agreement with experiment has been found by Wilson using calculations similar to those he used for the Λ-doubling in HF^+. Continuum wavefunctions were found to have a negligible contribution in this case (Table 3.16).

Table 3.16

Calculated and experimental spin–rotation constants in the $A\,^2\Sigma^+$ state of HF^+

	Calculated values (cm^{-1})			Experimental value (cm^{-1})
v	$\gamma_v^{(1)}$	$\gamma_v^{(2)}$	γ_v	γ_v
0	0.0005	0.5261	0.5266	0.5339
1	0.0004	0.5288	0.5292	0.5400
2	0.0004	0.5458	0.5462	0.5665

It is tempting to ask whether transitions are possible in either the interstellar or circumstellar media between the components of a spin doublet. Fig. 3.3 shows some of the energy levels in a $^2\Sigma^+$ state. Clearly the selection rules $+\not\leftrightarrow+$ $\;-\not\leftrightarrow-$ $\;+\leftrightarrow-$ make the transitions electric-dipole forbidden. However, the selection rules for quadrupole or magnetic dipole transitions are $+\leftrightarrow+$ $\;-\leftrightarrow-$ $\;+\not\leftrightarrow-$ $\;\Delta J = 0, \pm 1$ as required. We should expect these transitions to be much weaker (by a factor of several powers of ten) than ordinary electric-dipole allowed transitions. It is not, however, unreasonable that at a sufficient temperature for appreciable population of excited rotational levels, transitions between the spin-doublet components might be observed in space. Transitions between spin doublets have been observed terrestrially in O_2.

Spin-doubling in CaH has been studied by Cooper and good agreement has been obtained with experiment. In the limit of pure

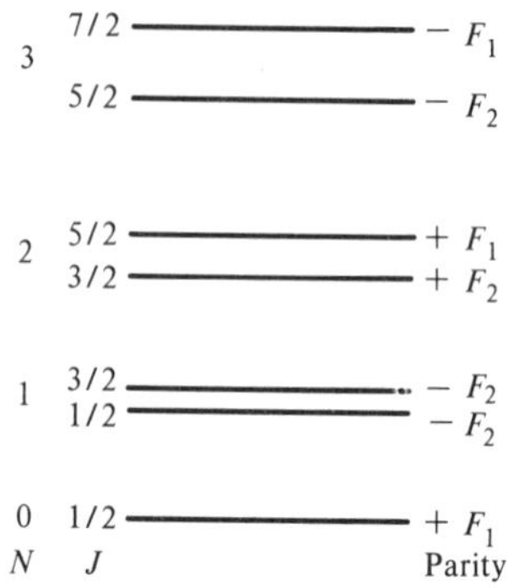

Fig. 3.3. Energy level diagram for a $^2\Sigma^+$ state showing spin splitting (not to scale). Splitting is given by $\gamma(N+\frac{1}{2})$.

precession, only $A^2\Pi$ is important in the summation for the spin–rotation coupling constant in the $X^2\Sigma^+$ state. Separate SCF procedures were used on each state and vibrational curves were constructed using Dunham coefficients generated from experimental data. The calculated value is about 13 per cent lower than experiment and it is necessary to include the contribution from the $E^2\Pi$ state. The contributions from the different states are labelled $\gamma_{X\text{-}A}$ and $\gamma_{X\text{-}E}$ in Table 3.17. The $L^2\Pi$ state is found to be unimportant. The final value obtained is within about 5 per cent of experiment.

Table 3.17
Spin-doubling in CaH ($X^2\Sigma^+$, $v''=0$)

		$\gamma_0^{(2)}/\text{cm}^{-1}$
Calculated	γ_{X-A}	0.03638
	γ_{X-E}	0.00771
Total		0.0441
Experiment (Berg *et al.*)		0.0424(10)

The figure in brackets is the quoted error in the last figure

The interactions of the orbital and spin angular momenta with the magnetic field may be described in terms of an effective g value. The component of the g value perpendicular to the molecular axis in a $^2\Sigma^+$ state may be written

$$g_{\perp}=g_e+4\sum_{n'v'}\frac{\langle {}^2\Sigma^+\ v|\mathrm{H_{SO}}|n'v'\rangle\langle n'v'|\mathrm{L}^+|{}^2\Sigma^+\ v\rangle}{E_{n'v'}-E_{\Sigma v}} \tag{3.8}$$

This is analogous to the expression for the spin–rotation constant, γ.

The e.s.r. spectrum of the $X^2\Sigma^+$ state of AlO undergoes anomalously large changes when the molecule is trapped in an inert gas matrix. Walker made a number of approximations in his calculation of $g_{\perp}$ for AlO: the summation was truncated after two low-lying ${}^2\Pi_i$ and ${}^2\Pi_r$ states had been included; all excited states were constructed from the virtual orbitals of the ground state; vibrational overlaps were approximated to unity. Despite these approximations, the model gave a very satisfactory qualitative explanation of the observed trend in $g_{\perp}$.

3.4. Perturbations in electronic spectra

We have already seem many examples of the importance of the mixing of electronic states through off-diagonal matrix elements of the spin–orbit coupling operator. The possibility of crossing from one state to another leads to a number of interesting effects in electronic spectroscopy. One manifestation of this is certain types of predissociation. A knowledge of the perturbation matrix element, together with the slopes of the two potential curves in the region of the perturbation, leads to an estimate of the crossing probability.

The operator $\mathrm{H_{SO}}$ couples not only pairs of states such as ${}^2\Pi$ and ${}^2\Sigma^+$, but also states such as ${}^2\Pi$ and ${}^4\Sigma^-$ (as discussed above in the case of NH^+). Consider, for example, the molecule BeO which has a ground state $X^1\Sigma^+$ and a low-lying excited state ${}^3\Pi$. Whereas the Wigmer–Witmer rules allow the ${}^3\Pi$ state to correlate with Be(^{1}S) + O(^{3}P), the ${}^1\Sigma^+$ state cannot dissociate to ground-state atoms. The question that arises concerns the magnitude of $\langle {}^1\Sigma^+|\mathrm{H_{SO}}|{}^3\Pi\rangle$. If there is sufficient mixing between ${}^3\Pi$ and ${}^1\Sigma^+$ then the ground state will produce ground-state atoms on dissociation. From the selection rules from non-zero matrix elements of $\mathrm{H_{SO}}$, we must calculate the interaction between $(1\text{–}4)\sigma^2 1\pi^4$ and $(1\text{–}4)\sigma^2 5\sigma 1\pi^+ 1\pi^- 1\bar{\pi}^-$. Stagg has calculated values of $\langle {}^1\Sigma^+|\mathrm{H_{SO}}|{}^3\Pi\rangle$ for the series of molecules BeO, MgO, and CaO and has in each case found that the matrix element is small. There will thus be a discrepancy between spectroscopic and thermodynamic dissociation energies.

Perturbations due to spin–orbit or electronic–rotational interactions are frequently expressed, as a first approximation, in terms of a single **a** parameter and a single **b** paramater

$$\mathbf{a} = \langle \pi | a l^{+} | \sigma \rangle \qquad\qquad \mathbf{b} = \langle \pi | l^{+} | \sigma \rangle \,. \tag{3.9}$$

A study of the series of molecules CO, CS, SiO, and SiS has produced values of **a** and **b** well within 10 per cent of experiment. One important conclusion to be drawn from this work is that a detailed interpretation of experimental results requires more refined *ab initio* calculations. Deperturbation studies may be improved by allowing different orbitals and different perturbation parameters for different pairs of interacting states. If we consider SiO, the $b^3\Pi_r$ perturbations lead to a value of **a** which varies from 7 cm^{-1} at 2.7 bohr to 130 cm^{-1} at 3.5 bohr. Nevertheless, these calculations are capable of giving some very detailed information. As an example we may cite the values of **a** calculated at the *R*-centroid of the $a^3\Sigma^+$ and $b^3\Pi$ states of these molecules: these *ab initio* calculations suggest why no $a^3\Sigma^+$–$b^3\Pi$ perturbations have been detected in experimental studies of SiO despite the fact that these perturbations are very important in CO and CS.

The alkali metal dimers are useful candidates for the study of isotope separation and for the construction of visible lasers. Off-diagonal matrix elements of H_{SO} are necessary for a quantitative understanding of some of the mechanisms that lead to dissociation. Fig. 3.4 shows that the $b^3\Pi_u$ state of Li_2 cannot decay by an allowed radiative transition but could dissociate into ground-state atoms via the repulsive $a^3\Sigma_u^+$ state which it crosses. H_{SO} couples $A^1\Sigma_u^+$ to $b^3\Pi_u$ and BJ^+L^- couples $b^3\Pi_u$ to $a^3\Sigma_u^+$. We thus have a possible mechanism for the dissociation of $A^1\Sigma_u^+$ via the $b^3\Pi_u$ state. Use of approximate spin–orbit matrix elements and uncertainty in the $b^3\Pi_u$ potential curve limits the reliability of calculations of predissociation lifetimes. Nevertheless, it is clear that crossing between curves can lead to the radiative destruction of specific rotation–vibration levels of the $A^1\Sigma_u^+$ state. Furthermore, this dissociation will be sensitive to the isotopic composition of the molecule.

3.5. Celestial masers

Both CH and OH are observed as masers in a number of interstellar objects. Indeed, CH is observed to mase in a very large range of

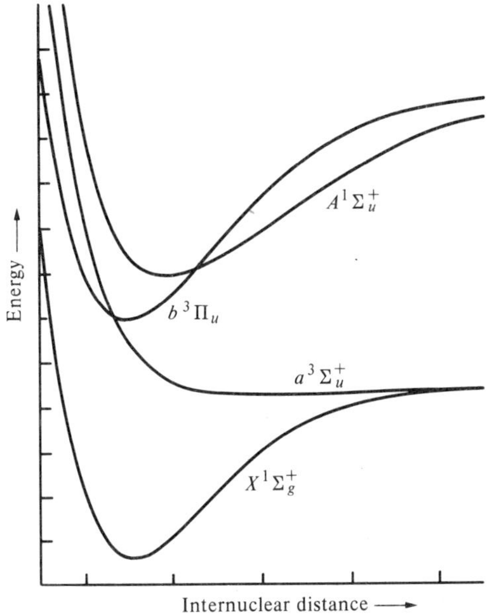

Fig. 3.4. Potential energy curves for diatomic lithium.

physical conditions: this would suggest a simple common cause for the observed population inversion. The two types of OH maser source are those associated with late-type stars (i.e. certain cool, red objects with oxygen-rich atmospheres – e.g. the M supergiant NML Cygnus) and those associated with H^+ regions. The most prominent H^+ region in the sky is the Orion nebula – the glowing gas being hydrogen. Behind this there is a cloud of H_2 containing clusters of infrared protostars and OH and H_2O masers. Study of these and other masers should eventually give a great deal of information about the velocities, temperatures and densities in protostars. It is interesting that the hydroxyl radical was the first interstellar molecule to be observed at radiofrequencies but that it is still probably the most difficult to understand.

Because the Boltzmann populations of Λ-doublet components are very nearly equal, very little population transfer would be required to the upper component of the Λ-doublet for maser action to occur. A model first suggested by Gwinn involves a collisional mechanism. Since spontaneous emission becomes more probable with increased energy separation, radiative decay from the collisionally populated state to the ground state is faster than the Λ-doublet transition.

There is thus a tendency for molecules to become trapped in the upper doublet component.

It is postulated that the transition states involved in the collisional process closely resemble the final state. If the π^+ and π^- orbitals are imagined as atomic p orbitals, then the possibility of partial bond formation stabilizes one transition state over the other. Fig. 3.5 shows collisions with the two Λ-doublet components of OH.

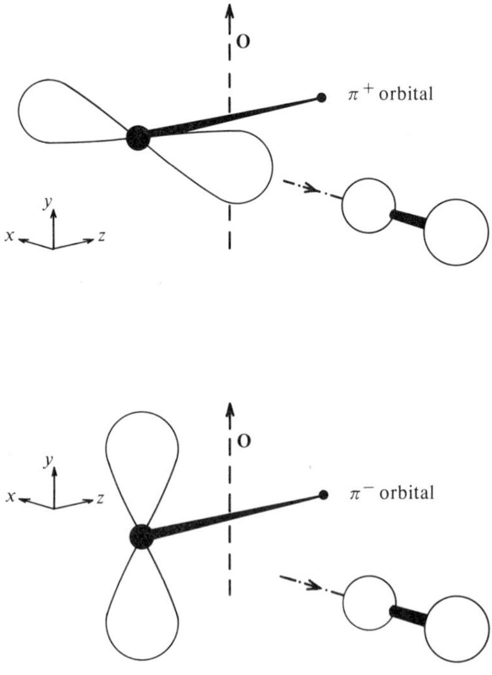

Fig. 3.5. Collisional interaction with the two Λ-doublet components of rotationally excited CH.

For a $^2\Pi$ state, the Λ-doublet component with the largest value of $\langle B\mathbf{J}.\mathbf{L}\rangle$ will be selectively populated for a π^1 electronic state. This situation is reversed for a π^3 electronic configuration. The calculation of matrix elements of $B\mathbf{J}.\mathbf{L}$ leads to expressions similar to those for p and q in Λ-doubling. The matrix elements of $B\mathbf{J}.\mathbf{L}$ show which Λ-doublet component in the excited state will be preferentially populated: electric-dipole allowed radiative transitions then lead to population inversion in the ground-state Λ-doublet. Fig. 3.6 shows part of the energy ladder for CH. The $B\mathbf{J}.\mathbf{L}$ matrix elements predict that the + component of the $J = 3/2$ state will be preferentially

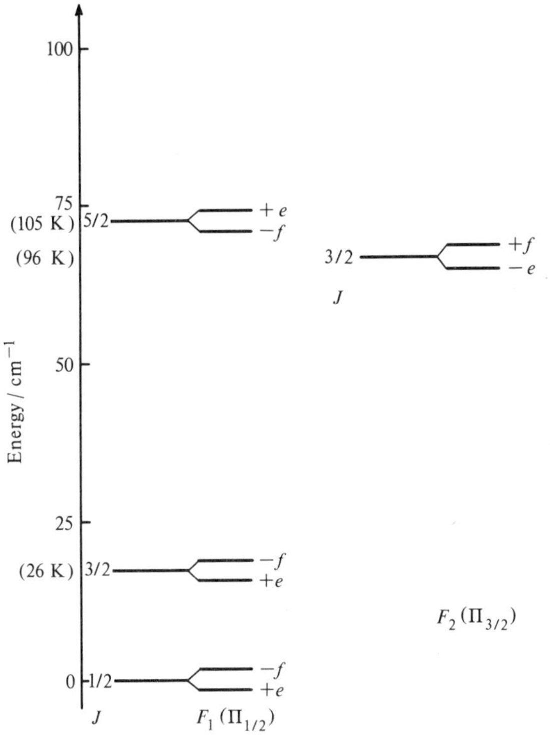

Fig. 3.6. Energy-level diagram for the low-lying rotational levels of CH.

populated and so radiative transition to the − component of $J = \frac{1}{2}$ leads to inversion. The results are consistent with observed maser action in CH and OH and can be used to predict the possibility of population inversion (caused by collisional excitation) in HF^+, HCl^+, and SH.

Maser action is still far from being completely understood. Returning to OH, some workers believe that this molecule might be produced with an inverted ground-state Λ-doublet by photodissociation of H_2O. On the other hand, both the $B\mathbf{J}.\mathbf{L}$ calculations and scattering calculations support a collisional pumping mechanism. For OH in circumstellar media, there is evidence for a pumping mechanism involving infrared radiation from the dust.

3.6. Summary

We have now seen a number of seemingly unrelated phenomena which are linked by dependence on off-diagonal matrix elements of the spin–orbit coupling operator. The results of calculations of Λ-doublet splittings are particularly pleasing. In the case of molecules close to the united-atom limit, very few states need be considered in the perturbation treatment. Indeed, the accuracy of the calculations for light hydrides is comparable to terrestrial experiment.

The expressions for g values in e.s.r. and spin–rotation constants, γ, are similar to the Λ-doubling expressions and valuable results have again been obtained. Some insight into interstellar maser action has been obtained by consideration of matrix elements of $B\mathbf{J}.\mathbf{L}$.

High precision microwave and radiofrequency measurements reveal deviations from the Mulliken and Christy formula even for low rotational levels. It is necessary to introduce higher order Λ-doubling parameters. *Ab initio* calculation of these should soon be possible.

Although at times only rather qualitative, deperturbation studies have produced important conclusions, particularly concerning predissociation; the use of accurate matrix elements to estimate predissociation lifetimes should be very important in the future.

3.7. Final conclusions

It would have been very easy to turn this short book into a very long one if we had extended its coverage to incorporate some of the many other aspects of molecular spin–orbit coupling which could legitimately be encompassed by our title. We chose to restrict our attention to the well-defined part of the topic where experimental measurements are accurate and unambiguous. These data not only challenge the theoretician but also provide an objective criterion for judgement on the success of calculations.

Within the realm of the diatomic molecule we have shown that computation of spin–orbit coupling constants is of comparable accuracy to that achieved for atoms. In the more interesting area of off-diagonal calculations of spin–orbit effects, the results are of similar accuracy to high-resolution spectroscopic studies. As such the work presented here denotes a definite break-point. Diatomic molecules are essentially a solved problem, despite the many specific instances which remain for research.

The careful work which has enabled this plateau to be reached should now provide the stimulus to go even further, and we hope that we have given the tools for others to attack the obvious extensions.

Our own preliminary work suggests that it should be possible to calculate spin–orbit splittings in small polyatomic species with *ab initio* methods. The wide areas of inorganic complexes and organic photochemistry then suggest themselves as the next challenges. All the preparative work for serious attacks on both topics has been provided.

So often the excuse for failing to extend accurate work on small molecules to larger systems has been the inadequacy of computer power. This barrier does not exist for the next steps in spin–orbit coupling in molecules. There are many research problems to be tackled in molecular spin–orbit coupling but no serious restrictions due to lack of theory in a form suitable for computation.

Further reading

1. Judd, B. R., *Angular momentum theory for diatomic molecules*. Academic Press, New York (1975).
2. Kovács, I., *Rotational structure in the spectra of diatomic molecules*. Adam-Hilger, London (1969).
3. Van Vleck, J. H., On σ-type doubling and electron spin in the spectra of diatomic molecules. *Phys. Rev.* **33**, 467 (1929).
4. Richards, W. G., Λ-doubling: a triumph for theory. *Chem. Britain* **15**, 68 (1979).
5. Hammersley, R. E. and Richards, W. G., Λ-doubling in the CH molecule. *Nature* **251**, 597 (1978).
6. Wilson, I. D. L., Λ-type doubling in the molecules $^{14}NH^+$, $^{15}NH^+$ and $^{14}NO^+$. *Mol. Phys.* **36**, 597 (1968).
7. Hall, J. A., Schamps, J., Robbe, J. M., and Lefèbvre-Brion, H., Theoretical study of the perturbation parameters in the $a^3\Pi$ and $A^1\Pi$ states of CO. *J. chem. Phys.* **59**, 3271 (1973).
8. Gold, E., Hammersley, R. E., and Richards, W. G., Possible new interstellar masers. *Proc. R. Soc. (Lond.) Ser. A* **373**, 269 (1980).

APPENDIX 3.A

THE OPERATOR R

In order to carry out a CI expansion of integrals such as $\langle {}^2\Pi | L^+ | {}^2\Sigma \rangle$ it is necessary to know the correct *relative* signs of all the determinantal-product states contributing to $|\nu\Lambda S\Sigma\rangle$ and to its projections of different Ω. For a given state, the Alchemy program outputs a list, in coded form, of the determinants being considered in the CI – these are only for the projection of highest Ω. The remaining determinants may be produced by successive operation of the spin-lowering operator (S^-) and a reflection operator (R) discussed by Judd.

We may summarize the effect of R by the rules:

(1) $R\psi_m = (-1)^m \psi_{-m}$

(2) $R\alpha = \beta \qquad R\beta = -\alpha$

(3) $R|\nu \ \Lambda \ S \ \Sigma\rangle = (-1)^{\Lambda+\Sigma-S} |\nu \ -\Lambda \ S \ -\Sigma\rangle$ (provided $\Lambda \neq 0$)

(4) If $\Lambda = 0$ we must include additional phase factors (+1) for Σ^+ and (−1) for Σ^- states.

Thus $R\sigma = \bar{\sigma} \quad R\bar{\pi}^+ = \pi^- \quad R\pi^- = -\bar{\pi}^+$ and so on.

We shall give an example of the use of R and shall check the result against that obtained with S^-.

Given

$$|{}^2\Sigma^-_{1/2}\rangle = \frac{1}{\sqrt{6}}|\sigma \ \pi^+ \ \bar{\pi}^-| + \frac{1}{\sqrt{6}}|\sigma \ \bar{\pi}^+ \ \pi^-| - \frac{2}{\sqrt{6}}|\bar{\sigma} \ \pi^+ \ \pi^-|,$$

then

$$R|^2\Sigma^-_{1/2}\rangle = -\frac{1}{\sqrt{6}}|\bar{\sigma}\ \bar{\pi}^-\ \pi^+| - \frac{1}{\sqrt{6}}|\bar{\sigma}\ \pi^-\ \bar{\pi}^+| + \frac{2}{\sqrt{6}}|\sigma\ \bar{\pi}^-\ \bar{\pi}^+|$$

$$= (-1)(-1)^{S-\Lambda-\Sigma}|^2\Sigma^-_{-1/2}\rangle$$

so

$$|^2\Sigma^-_{-1/2}\rangle = -\frac{1}{\sqrt{6}}|\bar{\sigma}\ \pi^+\ \bar{\pi}^-| - \frac{1}{\sqrt{6}}|\bar{\sigma}\ \bar{\pi}^+\ \pi^-| + \frac{2}{\sqrt{6}}|\sigma\ \bar{\pi}^+\ \bar{\pi}^-|.$$

Similarly

$$S^-|^2\Sigma^-_{1/2}\rangle = \frac{1}{\sqrt{6}}|\bar{\sigma}\ \pi^+\ \bar{\pi}^-| + \frac{1}{\sqrt{6}}|\sigma\ \bar{\pi}^+\ \bar{\pi}^-| + \frac{1}{\sqrt{6}}|\bar{\sigma}\ \bar{\pi}^+\ \pi^-|$$

$$+ \frac{1}{\sqrt{6}}|\sigma\ \bar{\pi}^+\ \bar{\pi}^-| - \frac{2}{\sqrt{6}}|\bar{\sigma}\ \bar{\pi}^+\ \pi^-| - \frac{2}{\sqrt{6}}|\bar{\sigma}\ \pi^+\ \bar{\pi}^-|$$

$$= -\frac{1}{\sqrt{6}}|\bar{\sigma}\ \pi^+\ \bar{\pi}^-| - \frac{1}{\sqrt{6}}|\bar{\sigma}\ \bar{\pi}^+\ \pi^-| + \frac{2}{\sqrt{6}}|\sigma\ \bar{\pi}^+\ \bar{\pi}^-|.$$

APPENDIX 3.B

SLATER'S RULES FOR OFF-DIAGONAL MATRIX ELEMENTS OF H_{SO}

Wilson has derived expressions for off-diagonal matrix elements of the spin–orbit operator between Slater determinants constructed from the same set of orthonormal orbitals.

Writing the spin–orbit operator as

$$H_{SO} = \sum_{\mu} (-1)^{\mu} \sum_{i} [h_i^{\mu} s_i^{-\mu} + \sum_{j \neq i} H_{ij}^{\mu} (s_i + 2s_j)^{-\mu}].$$

The case of $\mu = +1$ will be considered here and we shall write

$$\langle \pi^+ | \sigma \rangle = \frac{-1}{\sqrt{2}} \langle \pi^+(1) | h_1^{+1} | \sigma(1) \rangle$$

$$\langle ij | kl \rangle = \frac{-1}{\sqrt{2}} \langle i(1)j(2) | H_{12}^{+1} | k(1)l(2) \rangle.$$

(i) Two determinants differing by only one spin–orbital that is $\pi^+(\beta)$ in one and $\sigma(\alpha)$ in the other:

$$\langle \psi_1 | H_{SO} | \psi_2 \rangle = \langle p\pi^+ | q\sigma \rangle + \sum_{i\sigma\,(\alpha,\beta)} \langle p\pi^+ i\sigma | q\sigma i\sigma \rangle$$

$$- \sum_{i\sigma(\alpha)} [\langle p\pi^+ i\sigma | i\sigma q\sigma \rangle + 2\langle i\sigma p\pi^+ | q\sigma i\sigma \rangle]$$

$$- \sum_{i\sigma(\beta)} [2\langle p\pi^+ i\sigma | i\sigma q\sigma \rangle + \langle i\sigma p\pi^+ | q\sigma i\sigma \rangle]$$

$$+ \sum_{j\pi^+(\alpha,\beta)} [\langle p\pi^+ j\pi^+ | q\sigma j\pi^+ \rangle + 2\langle j\pi^+ p\pi^+ | j\pi^+ q\sigma \rangle]$$

$$-\sum_{j\pi^+(\alpha)} [\langle p\pi^+ j\pi^+ | j\pi^+ q\sigma\rangle + 2\langle j\pi^+ p\pi^+ | q\sigma j\pi^+\rangle]$$

$$-\sum_{j\pi^+(\beta)} [2\langle p\pi^+ j\pi^+ | j\pi^+ q\sigma\rangle + \langle j\pi^+ p\pi^+ | q\sigma j\pi^+\rangle]$$

$$+\sum_{j\pi^-(\alpha,\beta)} [\langle p\pi^+ j\pi^- | q\sigma j\pi^-\rangle + 2\langle j\pi^- p\pi^+ | j\pi^- q\sigma\rangle]$$

$$-\sum_{j\pi^-(\alpha)} [\langle p\pi^+ j\pi^- | j\pi^- q\sigma\rangle + 2\langle j\pi^- p\pi^+ | q\sigma j\pi^-\rangle]$$

$$-\sum_{j\pi^-(\beta)} [2\langle p\pi^+ j\pi^- | j\pi^- q\sigma\rangle + \langle j\pi^- p\pi^+ | q\sigma j\pi^-\rangle].$$

(ii) Two determinants differing by only one spin-orbital that is $\sigma(\alpha)$ in one $\pi^-(\beta)$ in the other

$$\langle\psi_1|\mathrm{H_{SO}}|\psi_2\rangle = \langle p\pi^+|q\sigma\rangle + \sum_{i\sigma(\alpha,\beta)} \langle p\pi^+ i\sigma | q\sigma i\sigma\rangle$$

$$-\sum_{i\sigma(\alpha)} [\langle i\sigma p\pi^+ | q\sigma i\sigma\rangle + 2\langle p\pi^+ i\sigma | i\sigma q\sigma\rangle]$$

$$-\sum_{i\sigma(\beta)} [2\langle i\sigma p\pi^+ | q\sigma i\sigma\rangle + \langle p\pi^+ i\sigma | i\sigma q\sigma\rangle]$$

$$+\sum_{j\pi^+(\alpha,\beta)} [\langle p\pi^+ j\pi^- | q\sigma j\pi^-\rangle + 2\langle j\pi^- p\pi^+ | j\pi^- q\sigma\rangle]$$

$$-\sum_{j\pi^+(\alpha)} [\langle j\pi^- p\pi^+ | q\sigma j\pi^-\rangle + 2\langle p\pi^+ j\pi^- | j\pi^- q\sigma\rangle]$$

$$-\sum_{j\pi^+(\beta)} [2\langle j\pi^- p\pi^+ | q\sigma j\pi^-\rangle + \langle p\pi^+ j\pi^- | j\pi^- q\sigma\rangle]$$

$$+\sum_{j\pi^-(\alpha,\beta)} [\langle p\pi^+ j\pi^+ | q\sigma j\pi^+\rangle + 2\langle j\pi^+ p\pi^+ | j\pi^+ q\sigma\rangle]$$

$$-\sum_{j\pi^-(\alpha)} [\langle j\pi^+ p\pi^+ | q\sigma j\pi^+\rangle + 2\langle p\pi^+ j\pi^+ | j\pi^+ q\sigma\rangle]$$

$$-\sum_{j\pi^-(\beta)} [2\langle j\pi^+ p\pi^+ | q\sigma j\pi^+\rangle + \langle p\pi^+ j\pi^+ | j\pi^+ q\sigma\rangle].$$

(iii) For matrix elements involving σ and π orbitals only, the following cases arise when the determinants differ by two spin-orbitals

(a) $\psi_1 = |\ldots p\sigma q\pi^+ \ldots|$ $\quad\quad$ $\psi_2 = |\ldots i\pi^+ j\pi^- \ldots|$

(b) $\psi_1 = |\ldots p\sigma q\pi^+ \ldots|$ $\quad\quad$ $\psi_2 = |\ldots i\sigma j\sigma \ldots|$

(c) $\psi_1 = |\ldots p\pi^+ q\pi^+ \ldots|$ $\quad\quad$ $\psi_2 = |\ldots i\sigma j\pi^+ \ldots|$

(d) $\psi_1 = |\ldots p\sigma q\sigma \ldots|$ $\quad\quad$ $\psi_2 = |\ldots i\sigma j\pi^- \ldots|$

(e) $\psi_1 = |\ldots p\pi^+ q\pi^- \ldots|$ $\quad\quad$ $\psi_2 = |\ldots i\sigma j\pi^- \ldots|$

(f) $\psi_1 = |\ldots p\sigma q\pi^- \ldots|$ $\quad\quad$ $\psi_2 = |\ldots i\pi^- j\pi^- \ldots|$

The spin functions for these orbitals must satisfy the condition that $\Delta\Sigma = -1$.

The matrix elements for each case are given by:

(a) $A\langle i\pi^- q\pi^+ | p\sigma j\pi^-\rangle - B\langle q\pi^+ i\pi^- | j\pi^- p\sigma\rangle$

$-C\langle j\pi^+ i\pi^- | p\sigma q\pi^-\rangle + D\langle q\pi^+ j\pi^+ | i\pi^+ p\sigma\rangle$

(b) $A\langle p\sigma q\pi^+ | i\sigma j\sigma\rangle + B\langle q\pi^+ p\sigma | j\sigma i\sigma\rangle$

$-C\langle p\sigma q\pi^+ | j\sigma i\sigma\rangle - D\langle q\pi^+ p\sigma | i\sigma j\sigma\rangle$

(c) $A\langle p\pi^+ j\pi^- | i\sigma q\pi^-\rangle + B\langle q\pi^+ p\pi^+ | j\pi^+ i\sigma\rangle$

$-C\langle p\pi^+ q\pi^+ | j\pi^+ i\sigma\rangle - D\langle q\pi^+ j\pi^- | i\sigma p\pi^-\rangle$

(d) $-\mathrm{A}\langle p\sigma j\pi^{+}|i\sigma q\sigma\rangle + \mathrm{B}\langle j\pi^{+}p\sigma|q\sigma i\sigma\rangle$

$-\mathrm{C}\langle j\pi^{+}q\sigma|p\sigma i\sigma\rangle + \mathrm{D}\langle q\sigma j\pi^{+}|i\sigma p\sigma\rangle$

(e) $\mathrm{A}\langle p\pi^{+}q\pi^{-}|i\sigma j\pi^{-}\rangle + \mathrm{B}\langle q\pi^{-}p\pi^{+}|j\pi^{-}i\sigma\rangle$

$-\mathrm{C}\langle p\pi^{+}q\pi^{-}|j\pi^{-}i\sigma\rangle - \mathrm{D}\langle q\pi^{-}p\pi^{+}|i\sigma j\pi^{-}\rangle$

(f) $\mathrm{A}\langle i\pi^{+}j\pi^{-}|p\sigma q\pi^{-}\rangle - \mathrm{B}\langle q\pi^{-}i\pi^{+}|j\pi^{-}p\sigma\rangle$

$-\mathrm{C}\langle j\pi^{+}i\pi^{-}|p\sigma q\pi^{-}\rangle + \mathrm{D}\langle q\pi^{-}j\pi^{-}|i\pi^{-}p\sigma\rangle.$

The functions A, B, C, and D are defined by

$$\mathrm{A} = \delta(\sigma_p, \sigma_j)\delta(\sigma_p, \sigma_i - 1) + 2\delta(\sigma_p, \sigma_i)\delta(\sigma_q, \sigma_j - 1).$$

$$\mathrm{B} = \delta(\sigma_p, \sigma_i)\delta(\sigma_q, \sigma_j - 1) + 2\delta(\sigma_q, \sigma_j)\delta(\sigma_p, \sigma_i - 1).$$

$$\mathrm{C} = \delta(\sigma_q, \sigma_i)\delta(\sigma_p, \sigma_j - 1) + 2\delta(\sigma_p, \sigma_j)\delta(\sigma_q, \sigma_i - 1).$$

$$\mathrm{D} = \delta(\sigma_p, \sigma_j)\delta(\sigma_q, \sigma_i - 1) + 2\delta(\sigma_q, \sigma_i)\delta(\sigma_p, \sigma_j - 1).$$

δ is the Kronecker delta function.

DETAILS OF THE MICROFICHE

In an envelope inside the back cover of this book will be found a microfiche containing listings of some of the spin–orbit routines. These are the routines for the calculation of one-electron and two-electron integrals of H_{SO}^{μ} (see Chapter 2) and off-diagonal matrix elements of H_{SO} and L^+S^- (see Chapter 3).

The integral routines are largely the work of Dr I. D. L. Wilson who is responsible for much of the work described in the second chapter. These routines form part of a suite of programs used for the calculation of diagonal matrix elements of the spin–orbit coupling operator, with or without configuration interaction. Both one- and two-centre components can be evaluated.

The off-diagonal matrix element program was written by Dr J. A. Hall. Much of the method is analogous to the diagonal case: the differences caused by the use of non-orthogonal orbitals are discussed in Chapter 3.

REFERENCES

Aarts, J. F. M., The Renner effect in $^2\Pi$ electronic states of linear triatomic molecules. I. Theory of vibronic interaction modified by spin-orbit coupling. *Mol. Phys.* **35**, 1785 (1978).

Abegg, P. W., *Ab initio* calculation of spin-orbit coupling constraints for Gaussian wave functions. *Mol. Phys.* **30**, 579 (1975).

Abegg, P. W. and Ha, T. K., *Ab initio* calculation of the spin-orbit coupling constant from gaussian lobe SCF molecular wavefunctions. *Mol. Phys.* **27**, 763 (1974).

Ackermann, F. and Miescher, E., Spin-orbit coupling in molecular Rydberg states of the NO molecule. *Chem. Phys. Lett.* **2**, 351 (1968).

Al-Mobarak, R. and Warren, K. D., Effect of covalency on the spin-orbit coupling constant. *Chem. Phys. Lett.* **21**, 513 (1973).

Aldeen, I. H. K., Allison, A. C., and Jamieson, M. J., Comparison of the methods for Λ-doubling. The $C^1\Pi_u$ states of H_2. *Mol. Phys.* **36**, 931 (1978).

Anholt, R., Meyerhof, W. E., and Salin, A., Effect of spin-orbit coupling on $^2P_{1/2}$-$^2P_{3/2}$ rotational transitions in heavy-ion collisions. *Phys. Rev.* **A16**, 951 (1977).

Barron, L. D. and Buckingham, A. D., Spin-orbit coupling and the interaction of molecules with the radiation field, *J. phys. B* **6**, 1295.

Bendazzoli, G. L. and Palmieri, P., Spin-orbit interaction in polyatomic molecules. *Int. J. quantum Chem.* **6**, 941 (1974).

Blume, M. and Watson, R. E., Theory of spin-orbit coupling in atoms. I. Derivation of the spin-orbit coupling constant. *Proc. R. Soc. (Lond.) Ser. A* **270**, 127 (1962).

Blume, M. and Watson, R. E., Theory of spin-orbit coupling in atoms. II. Comparison of theory with experiment. *Proc. R. Soc. (Lond.) Ser. A* **271**, 565 (1963).

Brand, J. C. D. and Stevens, C. G., Analysis of some singlet-triplet perturbations in the $^1A_2$ state of formaldehyde. *J. chem. Phys.* **58**, 3331 (1973).

Brown, J. M. and Watson, J. G. K., Spin-orbit and spin-rotation coupling in doublet states of diatomic molecules. *J. mol. Spectrosc.* **65**, 65 (1977).

Chatterjee, R. and Lulek, T., Origin of the spin-orbit interaction. *Acta Phys. Pol. A (Poland)* **A56**, 205 (1979).

Chatterjee, R., Dixon, J. M., and Lacroix, R., Comment on the relativistic correction term to the spin-orbit coupling. *Phys. Status Solidi B* **57**, K117 (1973).

Chiu, L.-Y.-C., Electron magnetic perturbation in diatomic molecules of Hund's case *b*. *J. chem. Phys.* **40**, 2276 (1964).

Chiu, Y.-N., Relativistic effect and non-conservative spin-orbital forces in the radiative transition of molecules. *J. chem. Phys.* **48**, 3476 (1968).

Christofferson, R. E. and Ruedenberg, K., Hybrid integrals over Slater type atomic orbitals. *J. chem. Phys.* **49**, 4285 (1968).

Cohen, J. S. and Schneider, B., Ground and excited states of diatomic neon and diatomic neon ion. I. Potential curves with and without spin-orbit coupling. *J. chem. Phys.* **61**, 3230 (1974).

Colbourn, E. A. and Wayne, F. D., The values of $\langle L^2 \rangle$ in diatomic molecules: implications for adiabatic and molecular fine structure calculations. *Mol. Phys.* **37**, 1755 (1979).

Cooper, D. L. and Richards, W. G., The accuracy of predicted radioastronomical frequencies and the spectrum of hydroxyl. *Nature* **278**, 624 (1978).

Coxon, J. A. and Hammersley, R. E., Spin-orbit coupling and Λ-type doubling in the ground state of hydroxyl. *J. mol. Spectrosc.* **58**, 29 (1975).

de Montgolfier, P. and Harriman, J. E., Study of different approximation in the calculation of *g* tensors in H_2^+. *J. chem. Phys.* **55**, 5262 (1971).

Dehmer, J. L., Phase-amplitude method in atomic physics. II. Z-dependence of spin-orbit coupling. *Phys. Rev.* **A7**, 4 (1973).

Deitrich, J., Multiconfiguration calculation of the oxygen ^3P ground state fine structure. *Phys. Rev.* **A11**, 1498 (1975).

Dousmanis, G. C., Sanders, T. M. and Townes, C. H., Microwave spectra of the free radicals OH and OD. *Phys. Rev.* **100**, 1735 (1955).

Dym, S. and Hochstrasser, R. M., Spin-orbit coupling and radiationless transition in aromatic ketones. *J. chem. Phys.* **51**, 2458 (1969).

Elliott, J. P., Theoretical studies in nuclear structure. V. The matrix elements of non-central forces with an application to the 2p-shell. *Proc. R. Soc. (Lond.) Ser. A* **218**, 345 (1953).

Ellis, R. L., Squire, R., and Jaffe, H. H., Use of the CNDO method in spectroscopy. V. Spin-orbit coupling. *J. chem. Phys.* **55**, 3499 (1971).

Ermler, W. C., Lee, Y. S., Pitzer, K. S., and Winter, N. W., *Ab initio* effective core potentials including relativistic effects. Potential curves for xenon diatomics. *J. chem. Phys.* **69**, 976 (1978).

Fano, U., Spin-orbit coupling: a weak force with conspicuous effects. *At. mol. Phys.* **2**, 30 (1970).

Fontana, P. R., Spin-orbit and spin-spin interactions in diatomic molecules. I. Fine structure of H_2. *Phys. Rev.* **125**, 220 (1961).

Gibbs, R. L., Two-electron spin-other-orbit integrals. *J. chem. Phys.* **51**, 5432 (1969).

Ginsburg, J. L. and Goodman, L., Effect of two electron spin-orbit interactions on singlet-triplet transition probabilities. *Mol. Phys.* **15**, 441 (1968).

Goodenough, J. B., Spin-orbit-coupling effects in transition-metal compounds. *Phys. Rev.* **171**, 466 (1968).

Green, S. and Zare, R. N., *Ab initio* calculation of the spin-rotation constant for $^2\Pi$ diatomics. *J. mol. Spectrosc.* **64**, 217 (1976).

Hall, J. A. and Richards, W. G., A theoretical study of the spectroscopic states of

the CF molecule. *Mol. Phys.* **23**, 331 (1972).
Hall, J. A., Schamps, J., Robbe, J. M., and Lefèbvre-Brion, H., Theoretical study of the perturbation parameters in the $a^3\Pi$ and $A^1\Pi$ states of CO. *J. chem. Phys.* **59**, 3271 (1973).
Hall, J. A., Walker, T. E. H., and Richards, W. G., One-centre matrix elements of the spin-orbit operator in linear molecules. Diagonal terms. *Mol. Phys.* **20**, 753 (1971).
Hall, W. R. and Hameka, H. F., Second-order effect of spin-orbit coupling on the angular dependence of zero-field splitting in CH_2. *J. chem. Phys.* **58**, 226 (1973).
Hammersley, R. E. and Richards, W. G., Λ-type doubling in the CH molecule. *Nature* **251**, 597 (1974).
Hammersley, R. E. and Richards, W. G., Λ-type doubling in the CD molecule. *Astrophys. J.* **194**, L61 (1974).
Hammersley, R. E. and Richards, W. G., *Ab initio* calculation of Λ-type doubling in excited rotational levels of the CH and CD molecules. *Astrophys. J.* **214**, 951 (1977).
Han, Pil Sun, Calculation of spin-spin interactions for zero field splitting in triplet biradicals. *J. Korean Phys. Soc.* **3**, 43 (1970).
Hay, P. J. and Dunning, T. H., The covalent and ionic states of the xenon halides. *J. chem. Phys.* **69**, 2209 (1978).
Hay, P. J., Wadt, W. R., Kahn, L. R., and Bobrowwicz, F. W., *Ab initio* studies of AuH, AuCl, HgH and $HgCl_2$ using relativistic effective core poentials. *J. chem. Phys.* **69**, 984 (1978).
Hellman, J. and Ballhausen, C. J., A calculation of the spin-orbit splitting of the $X^2\Pi$ state of NO. *Theoret. Chim. Acta* **3**, 159 (1965).
Henry, B. R. and Siebrand, W., Spin-orbit coupling in aromatic hydrocarbons. Analysis of non-radiative transitions between singlet and triplet states in benzene and naphthalene. *J. chem. Phys.* **54**, 1072 (1971).
Hermann, K. and Bagus, P. S., Hartree-Fock study of the interaction potential of He and Cl^+. *Chem. Phys. Lett.* **44**, 25 (1976).
Hinkley, R. K., Walker, T. E. H., and Richards, W. G., Spin-orbit coupling constants from Gaussian wave functions. *J. chem. Phys.* **52**, 5975 (1970).
Hinkley, R. K., Walker, T. E. H., and Richards, W. G., Spin-orbit coupling constants from semi-empirical wave functions. *Mol. Phys.* **24**, 1095 (1972).
Hochstrasser, R. M., Spin-orbit coupling in s-triazine. *Chem. Phys. Lett.* **17**, 1 (1972).
Horie, H., Spin-spin and spin-other-orbit interactions. *Prog. theoret. Phys.* **10**, 296 (1953).
Horsley, J. A. and Hall, J. A., Non-empirical calculation of spin-orbit coupling constants for some linear triatomic molecules. *Mol. Phys.* **25**, 438 (1973).
Huestis, D. L. and Schlotter, N. E., Diatomics-in-molecules potential surfaces for the triatomic rare gas halides. *J. chem. Phys.* **69**, 3100 (1978).
Ishiguro, E. and Kobori, M., Spin-orbit coupling constants in simple diatomic molecules. *J. Phys. Soc. Jpn.* **22**, 263 (1967).
Ito, H. and I'Haya, H. J., Evaluation of molecular integrals by gaussian expansion method with particular reference to zero field splitting integrals. *Chem.*

Phys. Lett. **17**, 516 (1972).

Ito, H. and I'Haya, Y., Evaluation of molecular spin-orbit integrals by a gaussian expansion method. *Mol. Phys.* **24**, 1103 (1972).

Jette, A. N., Fine structures of the metastable $C^3\Pi_u$ state of molecular hydrogen. *Chem. Phys. Lett.* **25**, 590 (1974).

Jungen, M., Spin-Bahn-Kopplungseffekte vershiedener Ordnung bei Jod und Dijodazetylen. *Theoret. chim. Acta* **27**, 33 (1972).

Kayama, K. and Baird, J. C., Spin-orbit effects and the fine structure in the ground state of O_2. *J. chem. Phys.* **46**, 2604 (1967).

Khachkuruzov, G. A. and Yurkov, G. N., Multiplet states of diatomic molecules with spin-orbital coupling intermediate between Hund's cases *a* and *b*. III. Dependence of the energy levels on the effect of molecular rotation and on the magnitude of the spin-orbit coupling. *Optics and Spectrosc.* **27**, 496 (1969).

Kovács, I., On the spin-orbit interaction in diatomic molecules I. *Can. J. Phys.* **36**, 309 (1958).

Kovács, I., On the spin-orbit interaction in diatomic molecules II. *Can. J. Phys.* **36**, 329 (1958).

Lambropoulos, P., Spin-orbit coupling and photoelectron polarization in multiphoton ionization of atoms. *Phys. Rev. Lett.* **30**, 413 (1973).

Lancelot, G., Theoretical and experimental studies on spin-orbit coupling in caffeine. *Mol. Phys.* **29**, 2099 (1975).

Langhoff, S. R., Spin-orbit contribution to the zero-field splitting in CH_2. *J. chem. Phys.* **61**, 3881 (1974).

Langhoff, S. R., Sink, M. L., Pritchard, R. H., Karn, C. W., Strickler, S. J., and Boyd, M. J., *Ab initio* study of perturbations between the $X^1\Sigma_g^+$ and $b^3\Sigma_g^-$ states of the C_2 molecule. *J. chem. Phys.* **67**,1051 (1977).

Leach, S., An application of population analysis to electronic spectroscopy: calculation of spin-orbit coupling constants for multiplet states of linear molecules. *Acta Phys. Polonica* **34**, 705 (1968).

Lee, S.-T., Suzer, S., and Shirley, D. A., Relativistic effects in the UV photoelectron spectra of group VI diatomic molecules. *Chem. Phys. Lett.* **41**, 25 (1976).

Lefèbvre-Brion, H. and Bessis, N., Spin-orbit splittings in $^2\Delta$ states of diatomic molecules. *Can. J. Phys.* **47**, 2727 (1969).

Lefèbvre-Brion, H. and Moser, C. M., Calculation os spin-orbit interaction in diatomic molecules. *J. chem. Phys.* **46**, 819 (1967).

Levy, P. M., Permutation degeneracy in the presence of large spin-orbit coupling. *Chem. Phys. Lett.* **3**, 556 (1969).

Lewis, W. B., Mann, J. B., Liberman, D. A. and Cromer, D. T., Calculation of spin-orbit coupling constants and other radial parameters for the actinide ions using relativistic wave functions. *J. chem. Phys.* **53**, 809 (1970).

Lin, K. C. and Lin, S. H., Theoretical calculation of an external heavy atom effect on the spin-orbit coupling of the benzene molecule. *Mol. Phys.* **21**, 1105 (1971).

Lin, S. H., Study of vibronic, spin-orbit and vibronic-spin-orbit couplings of formaldehyde with applications to radiative and non-radiative processes.

Proc. R. Soc. (Lond.) Ser. A **352**, 57 (1976).

Lindgren, B., On the spin-orbit coupling in the ground states of AsH and AsD. *Physica Scripta* **12**, 164 (1975).

Marvin, H. H., Mutual magnetic interactions of electrons. *Phys. Rev.* **71**, 102 (1947).

Masmanidis, C. A., Jaffe, H. H., and Ellis, R. L., Spin-orbit coupling in organic molecules. *J. phys. Chem.* **79**, 2052 (1975).

Matcha, R. L., Kern, C. W., and Schrader, D. M., Fine structure studies of diatomic molecules: two electron spin-spin and spin-orbit integrals. *J. chem. Phys.* **51**, 2152 (1969).

Matcha, R. L., Kouri, D. J., and Kern, C. W., Relativistic effects in diatomic molecules: evaluation of one-electron integrals. *J. chem. Phys.* **53**, 1052 (1970).

Matcha, R. N., Malli, G., and Milleur, M. B., Two-centre two-electron spin-spin and spin-orbit hybrid integrals. *J. chem. Phys.* **56**, 5982 (1972).

Meerts, W. L. and Dymanus, A., Accurate frequencies below 5 GHz of the lower J states of OD. *Astrophys. J.* **180**, L93 (1973).

Meerts, W. L. and Dymanus, A., The hyperfine Λ-doubling spectrum of sulfur hydride in the $^2\Pi_{3/2}$ state. *Astrophys. J.* **187**, L45 (1974).

Meerts, W. L. and Dymanus, A., A molecular beam electric resonance study of the hyperfine Λ-doubling spectrum of OH, OD, SH and SD. *Can. J. Phys.* **53**, 2123 (1975).

Meerts, W. L. and Dymanus, A., Electric dipole moments of OH and OD by molecular beam electric resonance. *Chem. Phys. Lett.* **23**, 45 (1973).

Meerts, W. L. and Dymanus, A., The hyperfine Λ-doubling spectrum of $^{14}N^{16}O$ and $^{15}N^{16}O$. *J. mol. Spectrosc.* **44**, 320 (1972).

Merer, A. J., Centrifugal distortion in the spin-orbit coupling of triplet states of light diatomic molecules. *Mol. Phys.* **23**, 309 (1972).

ter Meulen, J. J. and Dymanus, A., Beam-maser measurements of the ground-state transition frequencies of OH. *Astrophys. J.* **172** L21 (1972).

Michaels, H. H. and Hobbs, R. H., Electronic structure of the noble gas dimer ions – I. Potential energy curves and spectroscopic constants. *J. chem. Phys.* **69**, 5151 (1978).

Mijoule, C., Existence of a first order contribution of the spin-orbit coupling to the raising of degeneracy in zero field of triplet states in low symmetry molecules. *C.R. Hebd. Seances Acad. Sci. B* **278**, 857 (1974).

Miller, T. A., Zegarski, B. R., and Freund, R. S., Singlet-triplet anticrossings between $^1\Sigma_g^+$ and $^2\Sigma_g^+$ states of D_2. *J. mol. Spectrosc.* **69**, 199 (1978).

Mizushima, M., Molecular parameters of OH free radical. *Phys. Rev.* **A5**, 143 (1972).

Mulliken, R. S. and Christy, A., Λ-type doubling and electron configurations in diatomic molecules. *Phys. Rev.* **38**, 87 (1931).

Pritchard, R. H., Sink, M. L., and Kern, C. W., Theoretical study of the fine-structure coupling constants in the 2p $^3\Pi_u$ state of H_2. *Mol. Phys.* **30**, 1273 (1975).

Raftery, J. and Richards, W. G., Prediction of a predissociation in the $A^2\Sigma^+$ state of HCl^+ and DCl^+. *J. Phys. B* **6**, 1301 (1973).

Raftery, J. and Richards, W. G., Variation of the spin-orbit coupling in the molecular oxygen ion. *J. chem. Phys.* **62**, 3184 (1975).

Rajnak, K., Hartree-Fock calculation of many configurations of Uranium I. *Phys. Rev.* **A14**, 1979 (1976).

Rashev, S. and Plotnikov, V. G., Effect of n-electron delocalization on spin–orbit coupling of $n\pi^*$ and $\pi\pi^*$ states. *Opt. and Spectrosc.* **45**, 226 (1978).

Richards, W. G., Λ-doubling: a triumph for theory. *Chem. in Britain* **15**, 68 (1978).

Robbe, J. M. and Schamps, J., Calculations of perturbation parameters between valence states of CS. *J. chem. Phys.* **65**, 5420 (1976).

Roberts, P. J., Molecular four-centre integrals for tensor interelectronic interactions. *Proc. Phys. Soc.* **90**, 23 (1967).

Roche, A. L. and Lefebvre-Brion, H., Valence-shell states of PO. Variation of spin-orbit coupling constants with internuclear distance. *J. chem. Phys.* **73**, 1914 (1973).

Roothaan, C. C. J., Study of two-centre integrals useful in calculations on molecular structure. *J. chem. Phys.* **24**, 947 (1956).

Shimakura, N., Fujimura, Y., and Nakajima, T., Choice of basis sets for intersystem crossing. *Chem. Phys. Lett.* **40**, 222 (1976).

Sidis, V., Barat, M., and Dhuicq, D., Molecular orbital study of $Ar^+ + Ar$ collisions. *J. Phys. B* **8** 474 (1975).

Sidman, J. W., Spin–orbit coupling in the $^3A_2$–$^1A_1$ transition of formaldehyde. *J. chem. Phys.* **29**, 644 (1961).

Sommer, U., Torsion-dependent integrals for spin–orbit coupling in organic molecules. *Theoret. Chim. Acta* **9**, 26 (1967).

Sternheimer, R. M., Rodgers, J. E., and Das, T. P., Effect of the atomic core on the fine-structure splitting for excited nd and nf states of the alkali-metal atoms. *Phys. Rev.* **A17**, 505 (1978).

Stevens, C. G. and Brand, J. C. D., Angular momentum dependence of first- and second-order singlet-triplet interactions in polyatomic molecules. *J. chem. Phys.* **58**, 3324 (1973).

Stone, A. J., Spin-orbit coupling and the intersection of potential energy surfaces in polyatomic molecules. *Proc. R. Soc. (Lond.) Ser. A* **351**, 141 (1976).

Teague, M. R. and Lambropoulos, P., Three-photon ionization with spin–orbit coupling. *J. Phys. B* **9**, 1251 (1976).

Theis, W. R., A simple classical derivation of spin–orbit coupling. *Z. Phys. A* **290**, 355 (1979).

Thomas, L. K., The motion of the spinning electron. *Nature* **117**, 514 (1926).

Thorhallson, J., Fisk, C., and Fraga, S., Spin–orbit coupling in many-electron atoms. *J. chem. Phys.* **48**, 2925 (1968).

Trefftz, E., On the mutual influence of configuration interaction and spin–orbit coupling in heavy atoms. *J. Phys. B* **7**, L342 (1974).

Turro, N. J. and Devaquet, A., Chemiexcitation mechanisms. Role of symmetry and spin–orbit coupling in diradicals. *J. Amer. chem. Soc.* **97**, 3859 (1975).

Uhlenbeck, G. E. and Goudsmit, S., Spinning electrons and the structure of spectra. *Nature* **117**, 264 (1926).

Van Vleck, J. H., On σ-type doubling and electron spin in the spectra of diatomic molecules. *Phys. Rev.* **33**, 467 (1929).

Veseth, L., Modifications of the doublet energy formulae of a diatomic molecule necessitated by rotational stretching. *J. Phys. B* **3**, 1677 (1970).

Veseth, L., Spin-orbit and spin-other-orbit interaction in diatomic molecules. *Theoret chim. Acta* **18**, 368 (1970).

Veseth, L., Spin-rotation interaction in diatomic molecules. *J. Phys. B* **4**, 20 (1971).

Veseth, L., Corrections to the spin-orbit splitting in $^2\Pi$ states of diatomic molecules. *J. mol. Spectrosc.* **38** 228 (1971).

Veseth, L., Second-order spin-orbit splitting in $^2\Delta$ states of diatomic molecules. *Physica* **56**, 286 (1971).

Veseth, L., Fine structure and perturbations in the electronic spectra of HgH and HgD. *J. mol. Spectrosc.* **44**, 251 (1972).

Veseth, L., *Ab initio* calculation of a new spin-spin coupling constant for $^4\Pi$ states in diatomic molecules: the $a^4\Pi_u$ state in O_2^+ and $B^4\Pi$ state in VO. *J. mol. Spectrosc.* **77**, 154 (1979).

Veseth, L., On the Zeeman effect of $^2\Pi$ states in diatomic molecules. *J. mol. Spectrosc.* **77**, 195 (1979).

Wadt, W. R., The electronic states of Ar_2^+, Kr_2^+ and Xe_2^+. Potential curves with and without spin-orbit coupling. *J. chem. Phys.* **68**, 402 (1978).

Wadt, W. R. and Moomaw, W. R., The role of d-π orbitals in spin-orbit coupling. *Mol. Phys.* **25**, 1291 (1973).

Walker, T. E. H. and Barrow, R. F., The $A^2\Pi$-$X^2\Sigma^+$ system of BeF. *J. Phys. B* **2**, 102 (1969).

Walker, T. E. H., Berkowitz, J., Dehmar, J. L., and Waber, J. T., Non-statistical ratios of photoionization cross-sections for states split by spin-orbit coupling. *Phys. Rev. Lett.* **31**, 678 (1973).

Walker, T. E. H., Dehmar, P. M., and Berkowitz, J., Rotational band shapes in photoelectron spectroscopy: HF and DF. *J. chem. Phys.* **59**, 4292 (1973).

Walker, T. E. H. and Richards, W. G., *Ab initio* computation of spin-orbit coupling constants in diatomic molecules. *Symp. Faraday Soc.* **2**, 64 (1968).

Walker, T. E. H. and Richards, W. G., Assignment of molecular orbital configurations on the basis of Λ-type doubling. *Proc. Phys. Soc., (Lond.) (At. Mol. Phys.)* **3**, 271 (1970).

Walker, T. E. H. and Richards, W. G., Molecular spin-orbit coupling constants: role of core polarization. *J. chem. Phys.* **52**, 1311 (1970).

Walker, T. E. H. and Waber, J. T., Spin-orbit coupling and photoionization. *J. Phys. B* **7**, 674 (1974).

Wilson, I. D. L., Λ-type doubling in the molecules $^{14}NH^+$, $^{15}NH^+$ and $^{14}ND^+$. *Mol. Phys.* **36**, 597 (1978).

Wilson, I. D. L. and Richards, W. G., Λ-doubling in the SiH molecule. *Nature* **258**, 133 (1975).

Wilson, I. D. L. and Richards, W. G., Radioastronomical frequency for interstellar NH^+. *Nature* **271**, 137 (1978).

Wittel, K., Spin-orbit coupling in I_2^+. *Chem. Phys. Lett.* **15**, 555 (1972).

Yarunin, V. S. and Gavin, V. A., Spin-orbit coupling and the relative intensities of forbidden transitions in I_2 and IX molecules. *Opt. and Spectrosc.* **37**, 495 (1974).

Zaidi, H. and Verma, R. D., Quantum number dependence of the spin–orbit coupling in the $X^2\Pi$ state of PO. *Can. J. Phys.* **53**, 420 (1979).

Zamani-Khamiri, O. and Hameka, H. F., Spin–orbit contribution to the zero-field splitting of the oxygen molecule. *J. chem. Phys.* **55**, 2191 (1971).

INDEX